WHEN
TECHNOLOGY
WOUNDS

WHEN TECHNOLOGY WOUNDS

THE HUMAN CONSEQUENCES OF PROGRESS

CHELLIS GLENDINNING

William Morrow and Company, Inc.
New York

Grateful acknowledgment is made for permission to reprint the following:

Excerpt from *Woman and Nature,* by Susan Griffin. Copyright © 1978 by Susan Griffin. Reprinted by permission of Harper & Row, Publishers, Inc.

From *Knots,* by R. D. Laing. Copyright © 1970 by R. D. Laing. Reprinted by permission of Pantheon Books, a division of Random House, Inc.

"Manville Wins at the Bankruptcy Case" by Arthur Sharplin in *The AVA Advisor,* Volume 4, Number 3, Fall 1986. Reprinted by permission of Asbestos Victims of America/California and National Asbestos Information Center.

Recognizing the importance of preserving what has been written, it is the policy of William Morrow and Company, Inc., and its imprints and affiliates to have the books it publishes printed on acid-free paper, and we exert our best efforts to that end.

Library of Congress Cataloging-in-Publication Data

Glendinning, Chellis.
    When technology wounds: the human consequences of progress/
Chellis Glendinning.
        p.   cm.
    ISBN 0-688-07282-8
    1. Technology—Social aspects.   2. Health risk assessment.
I. Title.
T14.5.G58   1990
362.1'9698'00973—dc20   89-48093
                                                                          CIP

Printed in the United States of America

First Edition

1 2 3 4 5 6 7 8 9 10

BOOK DESIGN BY WILLIAM McCARTHY

For Jerry Mander

ACKNOWLEDGMENTS

Many people helped to make this book a reality. I am thankful to my editor and sidekick Shana Penn, who braved the snowbound New Mexico mountains in my International Scout and kept giving feedback long after her contract was complete. Michael Edelstein, Sarah Pirtle, and Craig Comstock read the manuscript and gave crucial commentary. Readers at earlier stages included Saul Landau, Rudolfo Mares, Ofer Zur, Marc Pilisuk, Tom Dalglish, Anne Valley-Fox, Philip Long, and Carole Roberts.

Gene Knudsen Hoffman, Wendy Grace, and Marc Kasky demonstrated unfailing loyalty to the project from start to finish. I am also grateful for the support of Barbara Hazard, Fridolin Smulders, Janet Kranzberg, Ann Dasburg, John Ross, the Henry Wright Estate, the Washington Research Institute, the San Francisco Firefighters, and the Fort Mason Foundation.

The research would not have been so thorough were it not for my colleagues who conducted interviews in faraway places: Wendy Miller, Liz Walker, Barbara Green, Susan Tieger, Carol Cotton, and, with enthusiasm above and beyond the call of duty, Marie Ferneau. Wayne Jacquith, Josephine Rohr, and Dorothy Lagarreta provided contacts for locating potential interviewees, and Neil Wollman gave professional feedback on research methodology.

For reliable sources of information I turned to the Foundation for the Advancement of Science and Education, M.J. Marvin of the Environmental Law Institute, Paul Scipione of the New Jersey Agent Orange Commission, Diana Hembre of the Center for Investigative Reporting, Steven Fox, American studies scholar at the University of New Mexico, and Tod Ensign of Citizen Soldier.

Throughout the project I received moral support from my friends Ann Lacy, Claire Greensfelder, Peter Barnes, Tamsin Taylor, and Marc Kasky. Lynn Richardson took time from a busy summer preparing for college to volunteer help with resource lists, and Jason Serinus hiked the Sangre de Cristo Mountains with me to challenge my thinking about the language of healing. As always, Susan Griffin came up with the title, with the help of Benina Berger Gould and Ed Schmookler.

I am profoundly thankful for the insights of those who inspire, inform, and give direction to my work: Jerry Mander, Langdon Winner, Michael Edelstein, Henry Vyner, Kai Erikson, Martha Wolfenstein, and, without equal, Lewis Mumford.

I am grateful to my publisher, James Landis, and my astute editor, Jane Meara. Thanks too to Dick Sugarman for legal consultation, Manon Pavy for all-round help, and David Falk for copyediting.

My deepest gratitude goes to the forty-six technology survivors who gave of themselves so courageously so that their stories could be told.

CONTENTS

The preservation of life
is the most urgent business
of the human race.

—Lewis Mumford, December 4, 1915

THE
STORIES

. . . the plight of things that have been created but not in a context of sufficient care.

—**Langdon Winner,** *Autonomous Technology*

MY STORY

I am a technology survivor. I became sick from an encounter with a health-threatening technology, and I survived. In 1967, when I was twenty years old, I went to a women's health clinic looking for advice about birth control. The nurses seated me in a classroom to hear a perfunctory lecture about condoms and diaphragms, but the attitude at the clinic unequivocally transmitted the message that women should use the "new technology": the Pill. No one advised me of possible harmful medical effects, and as the daughter of a physician raised to trust the dictates of the medical world, I never suspected any. As the sun set that afternoon, I found myself standing at the cashier's desk, my satchel loaded with packets of oral contraceptives.

I took them for two years. During that time I developed chronic vaginal infections, hives, food allergies, and paralyzing depression. Although no doctor was able to diagnose my condition at the time, I had contracted a disease called systemic candidiasis.

Then in 1971, after public hearings had revealed some of the Pill's more blatant medical effects, a gynecologist told me that "God's gift to woman" had arrived. It was the Dalkon Shield intrauterine device (IUD). He inserted one, and in 1973 I began to experience pain, fevers, and fatigue: pelvic inflammatory disease (PID) caused by the Shield. The antibiotics used to treat it worsened the candida problem until I suffered a total collapse.

Despite my efforts to get help, medical professionals did not seem to know that the root of my condition lay in immune dysfunction caused by ingesting artificial hormones and worsened by chronic inflammation. In all, my life was disrupted by illness for twenty years, including six years spent in bed. Healing did not begin until 1985 when a perceptive Chinese acupuncturist recognized the source of my ills. She had been confused by the assortment of symptoms I was presenting, and one day she decided to send me to a holistic doctor who specialized in diseases of immune dysfunction. Seeing him led me away, once and for all, from the Western medical technologies that had caused the illness and toward alternative approaches that catalyze the body's ability to heal. Finally and at long last, health began to flourish when I took treatment with homeopathic medicine.

For most of the years of illness, I lived in isolation with my problem. Doctors and the manufacturers of birth control technologies never acknowledged it or its source. Neither did my family, friends, or colleagues. I continually met with such incomplete and uncompassionate responses as: "What's your problem? You look fine to me!," "But your blood test is normal," and "What are you trying to teach yourself by giving yourself this illness?" There are many sources for the ignorance and denial that surround technology-induced disease, but the ultimate result of such frames of mind is that they alienate sick people from the support and caring they deserve and they alienate all of us from addressing the collective problem of technology's dangers.

I have written a book about technology survivors because I am concerned about the increasing threat modern technologies pose to our health and to the survival of the earth. I have written this book because I care about life.

FROM IUDS TO
ATOMIC BOMBS

CORRIDOR OF DEATH ALONG MISSISSIPPI: CHEMICALS
MAY BE KILLING THE OLD, THE YOUNG, THE UNBORN

—*San Francisco Examiner & Chronicle,*
  January 31, 1988

Gay and Patrick Ducey were working as janitors at a church in East
Oakland, California. It was 1970. A young couple, they had one
daughter and looked forward to having a big family. Twice that year
their task at the church was to apply weed killer to the grounds.
Suspecting that the chemicals could be poisonous, they took all the
precautions the literature of the day advised. They wore heavy clothing. Afterward, they scrubbed themselves, put their clothes in plastic
bags, and disposed of them at the dump. By the time a second
application was due, Gay was pregnant.

Shortly after the applications, Patrick developed migraine headaches and vertigo attacks, conditions doctors later diagnosed as
symptomatic of permanent neurological damage. Gay contracted a
chronic gastrointestinal ailment that disabled her for a year and for
which she still takes medication.

Seth was born with brain dysfunction, visual and hearing impairment, and a cleft palate. Their two cats died: one of a rare feline
cancer, the other after aborting two litters of deformed fetuses. Gay
and Patrick didn't understand why all these terrible things were
happening to them.

Then on Christmas Eve, 1978, while reading a *New Yorker* article
on the medical impact of pesticides, Patrick discovered the source of
their ills. The compounds 2,4-D and 2,4,5-T in the weed killer combined to form dioxin. Through this article and subsequent research,

17

Gay and Patrick learned that they had been exposed to an extremely deadly substance, that their medical ills mirrored those of other dioxin-exposed people, and that they had become, as they put it, "sitting ducks for cancer."

The story of the Ducey family is not unusual. In fact, such stories are becoming increasingly common. Millions of people in the world today have been directly exposed to toxic, health-threatening technologies in their workplaces, homes, food, medical treatments, and communities.

Love Canal. PCBs. The Dalkon Shield. Three Mile Island. Chlordane. Asbestos. Times Beach. Savannah River. Red dye No. 2. DES. At this point it is not unreasonable to suspect that in the United States, every single person has lived or worked in or near a hazardous situation. A partial list includes:

- 210,000–400,000 armed services veterans who witnessed nuclear testing in the South Pacific and Nevada[1]

- 9 million residents of Michigan potentially exposed to polybrominated biphenyl (PBB) flame retardant in 1973[2]

- 35 million oral contraceptive users[3]

- 21 million employees working with asbestos on the job[4]

- 16.2 million students and teachers each year working in schools built with asbestos[5]

- 1.5 million workers—uranium miners, lab workers, nuclear test site employees—working with radiation[6]

- 40 million residents of the Great Lakes region who potentially ingest toxic chemicals in drinking water and fish caught in the lakes[7]

- 30 million households treated with the pesticide chlordane since 1948[8]

- 7 million farm laborers, managers, and foremen working with pesticides on the job[9]

- 30 million households—and 96 million residents—living within fifty miles of a nuclear power plant[10]

- 100,000 workers at nuclear power plants exposed to measurable amounts of radiation each year[11]

- 14.6 million residents of New York City, Chicago, San Francisco, and Los Angeles exposed to electromagnetic radiation from the nation's most potent microwave sources[12]

- 50 million United States residents, most of them in rural areas, potentially at risk of exposure to pesticide-contaminated drinking water[13]

- 41.8 million Louisiana residents living along the industrial corridor from Baton Rouge to New Orleans where one fifth of the nation's petrochemicals are produced,[14] where 75 million pounds of industrial waste are dumped into the Mississippi River and there are 3,500,000 tons of toxic landfills[15]

- 8 million mothers, daughters, and sons exposed to the drug diethylstilbestrol (DES)[16]

- 250,000–500,000 Vietnam veterans exposed to herbicide Agent Orange[17]

- 2.2 million Dalkon Shield intrauterine device users[18]

- 120,000 residents living directly downwind of the Nevada Test Site where the nation's nuclear weapons are tested[19]

- 500,000 semiconductor production and electronics assembly workers working with toxic chemicals each year[20]

- 30 million video display terminal (VDT) users working around low-level electromagnetic radiation, microwave, ELF waves, and static electrical fields[21]

- 800,000 pregnant women prescribed medical X rays in one year[22]

- 1.2 million residents living near just 214 of the nation's 600,000 hazardous chemical dumps[23]

- 2,260,000 plastics employees working directly with vinyl chloride[24]

- 2 million workers working with benzene[25]

- 135 million residents living in 122 cities and counties charted below air quality standards for ozone and/or carbon monoxide levels[26]

- 240 million American citizens potentially exposed to 2.6 billion pounds of pesticides applied each year on crops, forests, lakes, city parks, lawns, playing fields, and in hospitals, schools, offices, and homes[27]

We live in a world of increasing numbers of health-threatening technologies—and increasing numbers of people made sick by technology. Today's development and use of technology pose danger not only to individual people like Gay and Patrick Ducey, but to life itself: to the essence and survival of the earth's waters and soil and air, to your life and mine.

The historian Lewis Mumford calls these times the Age of Progress, in which "the myth of the machine . . . has so captured the modern mind that no human sacrifice seems too great."[28] With the invention of the telephone, television, missiles, nuclear weapons, supercomputers, fiber optics, and superconductivity, the social system we inhabit has repeatedly favored technologies that usher us further and further away from the communal, nature-bound roots that for millennia honored life and interrelationship in human culture. In their place, the values fueling our modern concept of "progress" as unchecked technological development have become the moral imperative of the modern age—*and its curse.*

At this tenuous moment in history, then, meeting and befriending the survivors of health-threatening technologies can serve to awaken us to a pressing need: a comprehensive review of where modern technological society stands. In light of this need, the life experiences of those people who have become ill can no longer be confined to private reality. Revealed, they become a catalyst for opening our collective hearts to the passion and wisdom we need to make our world safe and livable. What the people who have endured the ordeal of technology-induced disease learn about technology, human relationships, and life's meaning are critical lessons for us all.

This book is about the lives and lessons of technology's survivors. While the inspiration to write the book comes from my own life, its content derives from a two-year research project.[29] The purpose of this research was to detail the psychological issues faced by people suffering from technology-induced disease. It was also to discover the ways they have invented for living meaningful lives after such disruption.

These themes had occupied my mind throughout my experience with medical technologies. They crystallized for me at the 1984 Radiation Survivors Congress in San Francisco. There I attended Dr. Henry Vyner's workshop on the psychological effects of radiation contamination. The participants were survivors of Hiroshima and Nagasaki, atomic veterans exposed to United States tests, downwinders from Utah, and nuclear test site workers. As I listened to their stories, my eyes widened. The psychological issues they were facing·were *the same* as those I encountered as a Pill/Dalkon Shield survivor. We shared the normal array of psychic suffering common to many people who have become seriously ill, but we also shared psychological repercussions that seemed unique, most notably: (1) a sense of helplessness owing to the physical fact of victimization; (2) a loss of social validation from people who have not questioned society's belief in technological progress; and (3) a shattering of faith in institutions we once believed in. In 1986, I launched an in-depth investigation of these issues, interviewing forty-six technology survivors sickened by an array of different technologies.

My goal in writing a book based on this research is, first and foremost, to speak to the thousands upon thousands of people who have endured illness caused by technology's excesses. Because we live in a society that so totally supports a particular kind of technological development, we live in a society that defends against the fact that its creations cause illness and death. My goal is to right the balance and in so doing, to deny the power of the machine to alienate us from our potential for caring. It is to give technology's survivors the recognition they deserve and to encourage them out of isolation into the community they already inhabit.

My second goal is to alert us all to the increasing dangers our technologies pose, from the very intimate dangers of an intrauterine device to the most pervasive and ultimate dangers of nuclear warfare. That our technologies so uniformly threaten health and life is no coincidence. Their development originates in a philosophical and political approach to life. This is an insight made by others before me, among them Lewis Mumford, Jacques Ellul, David Noble, Carolyn Merchant, Langdon Winner, Jerry Mander, and Susan Griffin. In the end, to silence the deathly knell our technologies increasingly sound requires us not merely to pass more stringent safety rules to regulate individual technics. It requires us to understand the philosophical and political underpinnings of the development and use of technology in our society—and to review our relationship to technology with that understanding in mind.

Perhaps the most astounding lesson I have learned while researching this book is that despite all the pain and suffering modern technologies inflict, in this age of technology the human spirit is alive and well. While I was researching and writing the book, many colleagues and friends took me aside and with concern in their voices, asked if I was doing all right. How could I, myself recovering from long-term illness, sustain optimism while spending so much time with people whose injured, malfunctioning, or deteriorating bodies dimmed personal hope? It is true that I encountered the darker chambers of the human heart, both in my new friends and in myself. I felt horror for what I saw them enduring and terror when I grasped the implications.

Yet the time I spent with these people often transcended any previous notion I had of healing. In the midst of so much uncertainty and pain, many of technology's survivors have taught themselves the capacity for real caring and joy. Having had belief and purpose wrenched from their lives, many have become creators of meaning and faith. All the survivors I interviewed had cultivated within themselves the spirit, the courage, and the creativity necessary to transform themselves from victims into something more like heroes, heroes of a technological age. This discovery holds promise for us all.

Before you meet this group of technology survivors, though, it is important to understand what I mean when I say "technology." The historian David Noble writes, "As technology has increasingly placed the world at people's fingertips, these people have become less able to put their fingers on precisely what technology is."[30] The insight has merit. In the eighteenth and nineteenth centuries, people spoke of machines, tools, crafts, factories, and industry as entirely separate things.[31] Today we use the term *technology* to describe a tool (hammer), a gadget (telephone answering machine), or a weapon (neutron bomb). It can name a technique (Swedish massage), an institution (ABC News), or a system (global network of multinational corporations), and it can describe all of these things at once.

A broad definition of technology is in order. We can agree with Lewis Mumford who says that technology includes those instruments that accomplish work, plus the skills and methods we need to use them.[32] Political scientist Langdon Winner adds to this definition when he points to mechanistically organized networks and systems like governments, armies, and bureaucracies.[33] Mumford and, later, Noble go one step further. They include people. Without specially trained participants performing the tasks of technological organizations—participants like assembly line workers, administrators, and research and development teams—they cannot function.[34]

Such different kinds of things can be linked under one term because they share common qualities. First, all of them—from hammers to boards of directors—perform work some person has deemed important. Second, they are rational and efficient in their approach to the work. Last, they emanate directly from the field of science, a philosophical approach that views all of life, including people, as extensions of one great machine called the universe.

I include chemicals, radiation, and biotechnologies in my definition. If technology is defined by a mechanistic approach, the drive to efficiency, and the ability to perform work, a modern definition must include the catalyzers of material transformations. In earlier times these were the substances used to brew, tan, and distill. The original biotechnologies were the yeasts and bacteria used to ferment alcohol and manufacture cheese. In today's world these catalyzers have become the ingredients and products created when scientists fragment natural substances like plant leaves, asbestos, and atoms, or when they alter their chemical or genetic composition into dioxin, diethylstilbestrol, and recombinant rabies—substances that then go on to perform some form of work.

I decided to limit what technologies I would choose for my research. I did not seek out people made sick by technical organizations or networks, although there is a case to be made for a causal relationship between many of these arrangements and human illness. Rather I interviewed people affected by actual machines, tools, and chemicals. I do draw a line in my choice of such technologies. I do not include common household appliances like electric heaters and toaster ovens, although these machines have done their share to harm people. I include technologies that directly expose people to illness or death on a large scale, not by accident, but because they are inherently dangerous. Damage to human health is not haphazard. It is a foregone conclusion. I include pesticides and herbicides, ionizing and nonionizing radiation, intrauterine devices, artificial hormones, food additives, toxic waste, industrial chemicals, and toxic building materials. These are often the technologies that have been employed without adequate assessment of human or environmental impact, without democratic decision-making about their use, and more often than not, without the user's knowledge of their danger.

This book focuses on the psychological impact of technology-induced illness from not one, but many modern technologies. I make this choice because whether the agent of illness is a pill or a pesticide, the psychological issues one faces are similar. This is a new insight. To date, mental health researchers have only begun to make conclusive generalizations about the psychological effects of technology-induced disease.[35] One reason for this dearth is, predictably, the result of the technologization of education. Specialization has so fragmented academia that researchers focus on ever more minute aspects of reality, and few see the whole. Psychological researchers studying the effects of catastrophe and disaster tend to focus on singular events like incest or tornadoes and not to make generalizations, as psychiatrist Ronnie Bulman-Janoff points out, "across victimizations."[36]

A second reason for a lack of overview is the tremendous variation among technological events. Having an IUD inserted into one's uterus is an experience qualitatively different from witnessing a thermonuclear blast. Some technological events seem normal, within the range of everyday life, while others are blatantly catastrophic.

Then there are variations in what kinds of disease a person develops. For some, it is an illness that may be taxing but ultimately can be healed. For others, the resulting illness is grueling and chronic. For still others, it is terminal. Some technology-induced diseases appear the moment of exposure, while others evolve over twenty years' time. Each illness, with its particular demands and implications, can contribute to a different kind of psychological response.

There are also differences in the individuals affected: in their personalities, coping styles, and belief systems. One person conscientiously suppresses what has occurred and attempts to move on. Another is emotionally consumed by each medical, legal, and psychological development. One person believes that the government is evil for perpetrating a particular technology, another that he attracted the contamination to himself.

Then there are variations of the historical moment a technological event alters a person's life. For some, it occurs when they have strong psychic, physical, and social reserves. For others, the event is the last straw in a series of overwhelming life experiences. Some people have a history of trial and hardship. Others do not.

Last, there are cultural variations that can frame a person's response to technology-induced disease. Large Hispanic families may band together to help a relative, and Spanish culture may offer spiritual sustenance. On the other hand, the relative isolation of suburban life may make coping difficult. Variations in economic reserves either offer possible opportunities for effective coping or else leave a person bankrupt on every level. Many differences militate against the possibility of generalizing about the psychological effects of technology-induced disease.

Ultimately, though, the practice of looking at "victimizations" separately has a predictable effect: It maximizes the likelihood of finding differences and minimizes the possibility of seeing commonalities, making a vision of the whole picture and all its interconnections impossible. Yet, in the end, the common psychological issues and responses shared by technology survivors cannot be overlooked. My initial insight at the National Radiation Survivors Congress has echoed throughout the preparation of this book. In my encounters with survivors, I have found myself marveling: *"It's the same story over and over again."* I want to tell that story.

THE PEOPLE

And, indeed, these one-eyed prophesies came true.

—Lewis Mumford, *My Works and Days*

The overall story of technology's wounds is made up of the individual stories of hundreds of thousands of survivors. These are the people whose lives have been altered by an encounter with a health-threatening technology. The following are the forty-six who told me about their lives. For personal reasons, some of them have requested that their real names not be used.

Sukey Fox was living in a small town in Colorado when the Dalkon Shield intrauterine device was prescribed for her.

Beryl Landau is a California artist who used oral contraceptives and the Dalkon Shield IUD.

Roberto Garcia worked as an electrician and a pipe cover worker at Mare Island Naval Shipyard in California. He was exposed to asbestos.

Perry Styles is a DES daughter from Massachusetts.

Lawyer Robert McIntyre was first exposed to mercury in his dental fillings and then to various pesticides, including paraquat that had drifted onto his Texas ranch from a nearby grape vineyard.

Harriet Beinfield lived near a national forest in northern California where pesticides were routinely sprayed. She also took birth control pills and wore a Dalkon Shield IUD.

Roy Kimura, an American living in Japan during World War II, was exposed to radiation when he went to Hiroshima to find relatives after the bombing.

Rhiane Levy became sensitive to the electromagnetic radiation emanating from computers, fluorescent lights, airplanes, automobiles, and TV stations.

Meryl Tavich took oral contraceptives.

Californian Pat Cody was prescribed DES in the 1950s.

Laura Martin-Buhler was living in Utah during the aboveground nuclear tests in Nevada.

June Casey was going to college downwind of Hanford Nuclear Reservation in Washington State when the complex released iodine 131 into the atmosphere.

Benina Berger Gould was prescribed birth control pills, the Copper-7 and Dalkon Shield IUDs, and estrogen replacement therapy.

Susan Griffin used the Pill, the Dalkon Shield, antibiotics, and cortisone drugs.

Californian Carey Wallace used the Dalkon Shield.

Charlotte Mock drank polluted water from her faucet in Albuquerque, New Mexico.

Gay Ducey was living in Oakland, California, when she applied a common weed-killer to her church lawn. She was pregnant at the time.

Lois Gibbs lived in Love Canal, New York.

Marmika Paskiewicz was working as a museum exhibition specialist in Santa Fe, New Mexico, when she was exposed to toxic art materials, including xylene and 1,1,1-trichloroethane.

Loran Calvert was a machinist and test specialist in Vallejo, California, exposed to asbestos on the job.

Jane Woolf was caretaking a farm in Michigan when the water became contaminated with benzene from a leak at a nearby gas station.

Susan Hernandez worked at the GTE Lenkurt electronics plant in Albuquerque, New Mexico, where she was exposed to dozens of toxic chemicals.

Diane Carter was given a Dalkon Shield IUD in Denver, Colorado.

Genevieve Hollander received one at a clinic on Long Island.

Jesus Rives worked as a tool mechanic in California. He was exposed to asbestos.

Jose Luis Roybal was sent to guard Hiroshima in 1945 after the bombing.

Gilberto Quintana was given shore leave from a U.S. Navy ship in the fall of 1945 in Nagasaki.

Kari Pratt was exposed to various pesticides drifting from a vineyard in Texas. She was also exposed to the pesticide lindane sprayed in her home in Iowa.

Hydrologist Teddy Ostrow was exposed to multiple toxic chemicals on the job in hazardous waste management in California.

Oklahoma artist Marie Ferneau was exposed to cyanide fumes in her sculpture studio.

Minnesotan Anika Jans used birth control pills and the Dalkon Shield and was overprescribed broad-spectrum antibiotics.

Working as a cook in Washington State, Debbie La Monica was exposed to toxic chemicals in various kitchen products.

The Army sent Ricardo Candelaria to Yucca Flats, Nevada, to witness an atomic blast.

George Milne saw nuclear blasts at the United States test sites in the South Pacific.

Nathan Robinson was a sheet metal worker in New York and California. He was exposed to asbestos.

Irene Baca worked at the GTE plant in Albuquerque.

Wendy Grace was prescribed a Saf-T-Coil IUD in New York City.

Betsy Berning was given a Dalkon Shield in Cincinnati, Ohio. She was also prescribed seven years of antibiotic treatment.

Heather Maurer worked in her father's plumbing business where she was exposed to asbestos. She also wore a Dalkon Shield IUD.

Sarah Pirtle is a Massachusetts DES daughter.

Bliss Bruen was exposed to chlordane during a routine regulation spraying of her home in Washington, D.C.

While attending Harvard Law School, Danny Morris was exposed to pesticides sprayed in his neighborhood.

Andy Hawkinson was exposed to radiation fallout at Eniwetok Island in the South Pacific, where he was serving with the U.S. Marines.

Thérèse Khalsa used diet foods spiked with aspartame.

Maria Carriaga was living next door to a particle board factory in New Mexico when she became exposed to formaldehyde and toluene.

Over a period of years, Carl Porter was exposed to toxic anesthesia gases in the operating room of a San Francisco hospital.

THE STORIES

The weight of this sad time we must obey:
Speak what we feel, not what we ought to say.

—**William Shakespeare,** *King Lear*

Meet Andy Hawkinson, Thérèse Khalsa, Maria Carriaga, and Carl Porter. Their stories stand as testimony that such an experience can happen to anyone, and it can happen anywhere—in the Army, the kitchen, the neighborhood, or the hospital. Their stories are testimony that such an encounter can have far-reaching and irreversible effects.

In 1957, eighteen-year-old Andy Hawkinson was a military policeman for the United States Army. A short and brawny man, he was energetic about life, enthusiastic about the Army, and proud of his role in defending his country. That year he was assigned to Eniwetok Atoll in the Marshall Islands as a custodial soldier. He spent eight months there, roaming the mile-long tropical island and flying patrol to neighboring atolls.

To Andy, the technological event that shaped his life was a nonevent. "One of the definitions of an event is that it is perceived," he explains. "My problem is that what contaminated me was invisible. Eniwetok seemed idyllic. Beautiful blue lagoon. Sunshine. Constant warm breeze. There was swimming in the lagoon, speedboats, fishing. And all I had to do was—eight hours a day, five days a week—ride around with a partner and take a few drunks back to their barracks. To this day I can honestly tell you that Eniwetok was an environmental paradise because when I was there, it was paradise to me. I knew no better."

At the time, though, Andy did wonder. World War II and the Korean War were both over. He could not figure out why he, the

29

U.S. Coast Guard, the Army, and the Air Force were in the South Pacific. Despite his questions to superior officers, no one would give him an answer.

Then unexpectedly, Andy's father died, and he returned to the States for the funeral. In those days the draft was supplying the armed services with adequate numbers of young men for overseas duty, and so the Army restationed Andy, an only son, near his mother's home at Fort Sheridan, Illinois.

After he left Eniwetok, twenty-one atomic and hydrogen bombs were exploded there and on neighboring islands. Years later Andy learned that before he had arrived, between 1948 and 1956, twenty-two bombs had been exploded there, including Edward Teller's infamous fifteen-megaton hydrogen bomb. During his stay in the South Pacific, Andy says, "I lived at ground zero."

But he didn't know this at the time—the Army never told him—and he didn't find out for twenty years. Health problems led to the discovery. In 1977, Andy was married, had a family, and was working as a manager of a consumer finance office in San Jose, California. Suddenly and unexpectedly he developed cataracts. This was unusual because the condition ordinarily evolves over time and normally affects older people. Andy had two surgeries, one that year and one the following year, and they were successful. With contact lenses his sight was restored. All was seemingly well.

Then three months after the last surgery, as Andy describes it, "One Sunday morning I got up to read the newspaper, and something was wrong. I couldn't see the print. It was like you take a pie and quarter it, and there was a big piece missing. That was the beginning. The retina had separated. I had to rush to the hospital and have immediate surgery because if you don't, the optic nerve will die. They sewed a silicone band around my eye to hold the retina in place, and it was the most painful experience of my life." Five more surgeries followed, each unexpected, each anguishing, expensive, and disruptive.

As these events unfolded, Andy was puzzled. He was a young man, thirty-eight years old, in otherwise good health. He began to spend his lunch hours poring over medical journals for clues, but he came up with no explanations. Then *People* magazine published an article about two women who had lost their husbands. Both men had been on Eniwetok. Their wives were now trying to get compensation from the government for wrongful death. This discovery set Andy wondering. Next he came across a newspaper article featuring a man named Orville Kelly. Kelly had also been at Eniwetok. He had health problems too,[1] some similar to Andy's, and he had started an organization called the National Association of Atomic Veterans (NAAV). Andy finally understood that he had suffered because of exposure to nuclear testing. He was relieved to know. He was also terrified.

The United States government began nuclear testing in 1945 when it exploded the first atomic bomb at Alamogordo, New Mexico. Since then it has set off over 900 bombs in the South Pacific, at the Nevada Test Site, and at various lesser-known locations like San Diego; Hattiesburg, Mississippi; Rifle, Colorado; and Amchitka, Alaska.[2] One purpose of these tests was to expose people directly to an explosion to see if they could function afterward. How close was too close? Could a foot soldier in a foxhole muster the strength to fight after a blast? Could a pilot in a nearby plane keep flying? Could a sailor keep working on a ship? Would protective equipment help?

From 1946 to 1962, the government ordered over 210,000 servicemen and servicewomen to witness hundreds of nuclear explosions.[3] Some were ordered to sit in trenches one or two miles from detonation and then march directly to ground zero to execute military maneuvers. Some were stationed on ships that steamed their way toward the mushroom cloud, while others flew airplanes right through the cloud. Andy was one of those servicemen. During that time, over 250,000 civilians also worked at the various sites.[4] And through the years United States testing programs have also contaminated unidentified numbers of American citizens—from the documented downwinders of St. George, Utah, to the millions of residents living in line of wind patterns carrying fallout.

Radiation particles rip through the structure of human cells, leaving them either dead, or damaged and open to disease.[5] Attendance at any convention of NAAV or the newer National Association of Radiation Survivors (NARS) reveals a startling concentration of people struggling with deteriorating health, from cataracts to chronic fatigue immune dysfunction syndrome to cancer. Specifically regarding the impact of nuclear testing, Dr. Lester Lave and his colleagues at Carnegie-Mellon University have found that increases in the number of infant deaths in the 1960s directly correlate with the timing of nuclear tests,[6] and in a 1984 study of the health effects of Nevada testing, Dr. Carl Johnson reveals a significant increase in leukemia, lymphoma, thyroid cancer, breast cancer, melanoma, bone cancer, and brain tumors downwind of the test site.[7]

Meanwhile, nuclear testing continues in this country, the Soviet Union, China, and the South Pacific. The 1963 Partial Test Ban Treaty sent most testing underground, and many people believe that such testing is safe. In fact, though, underground explosions do release radiation. The much, and rightly, feared Chernobyl accident offers a revealing comparison. It released some 43 million curies of iodine 131 and cesium 137. According to Colonel Raymond Brim, in charge of fallout monitoring for the Air Force, until 1979 United States underground tests had leaked *62 million curies.*[8] This figure does not include the accelerated testing schedules of the Reagan and Bush administrations after 1981.

Since his eye surgeries, Andy has learned to adjust to visual limitation. He wears glasses that magnify the immediate environment three times, and he doesn't drive. Although he can see somewhat and even read, Andy is legally blind. He has suffered three bouts of internal bleeding, each sending him to the hospital and once requiring thirteen pints of blood to keep him alive. He has sustained two heart attacks.

Partially blind, slowed down by heart disease, and wary of future emergencies, Andy is against nuclear testing and weaponry, technologies that he says have caused "devastation to mankind and my own pain and suffering." At forty-seven, he says, "I really don't think I'll be alive another five years."

Thérèse Khalsa was on the front lines elsewhere—in her kitchen in Santa Fe, New Mexico. At age thirty-seven, like many other people approaching the middle years, she was on a diet. Unlike many people, though, Thérèse was extremely conscious of the value of nutrition and preventive health care. She had practiced yoga for years, and as a chiropractic assistant she received regular chiropractic and massage treatments. Her diet consisted of all-natural, usually organic foods. Since she was trying to lose weight, she included in it a low-calorie sweetener that was touted in advertisements as natural. The sweetener was NutraSweet, a chemical compound called aspartame. From September 1986 to January 1987, Thérèse drank diet colas and ate cereals containing aspartame, and she occasionally added NutraSweet to a protein drink she had for breakfast. The diet was going well.

Then in December, a few days before Christmas, Thérèse had her first "attack." She had received disappointing news from her brother, and feeling upset, she left the house for the natural foods store. Once in the store, Thérèse was overcome with anxiety. As she describes it, "I didn't know who I was, where I was, or what I was doing. I felt strange inside, dizzy, nauseated, and confused. It was like people probably feel when they take drugs. The aisles appeared to extend up on either side, and I wondered if I was walking straight. My heart was pounding. Finally I just sat down. I hardly dared to move. I hardly dared to breathe. The store manager drove me home."

The experience lasted an hour. It was repeated again and again that week, lasting one to several hours each time. Thérèse became incapacitated, afraid to perform her job, take care of her children, drive a car, or even walk outside. She sat at home, leaving only for doctor's appointments.

After a week a general consensus was developing among her doctors and family: Thérèse was under a great deal of stress and was having psychologically induced anxiety attacks. Christmas was around the corner, and the patient load at work seemed to increase with the holiday season. Her children needed to be driven to and from countless parties and classes. Thérèse blamed herself for being a "perfectionist" and pushing herself to control the myriad responsibilities in her life.

In January, Thérèse took a week off and went to a mental health clinic in Milwaukee for rest and psychological counseling. Despite the help she received, the "attacks" continued as before, but now with a new twist. Every night Thérèse had a seizure in which her body shook uncontrollably, her heart pounded, and her teeth chattered.

Then by happenstance, Thérèse discovered what she believes to be the source of the problem. For Christmas she had given a friend a special gift: a bottle of vitamin C sweetened with aspartame. In mid-January the friend reported back to Thérèse: "When I take these pills, I immediately feel dizzy." They wondered if it could be the aspartame, and they began to research.

The sugary-tasting aspartame is a small molecule comprised of several components, including phenylalanine, aspartic acid, and methanol. In fact, phenylalanine and aspartic acid have natural analogues. They are amino acids, but ordinarily in nature they combine with as many as eighteen other amino acids to form protein. Because aspartame consists of only these two, it delivers them in a much more concentrated form than a person would normally consume.[9] At this point, scientific studies on the benefits or dangers of aspartame contradict one another—with those made by independent medical researchers revealing health dangers for humans and animals, and those by the Food and Drug Administration (FDA) and the sweetener's manufacturer, Monsanto Chemical Company, refuting health dangers. As of July 1989, though, the FDA had received more than forty-five hundred complaints listing some seventy symptoms, among them blindness, seizures, blackouts, coma, memory loss, headaches, heart palpitations, and menstrual irregularities.[10]

According to Dr. Harvey Levy of the Harvard Medical School and Dr. Louis Elsas of Emory University, the phenylalanine in aspartame can cause irreversible brain damage to a growing fetus or an infant, and particularly to the children of women who are genetic carriers of a metabolic disorder called phenylketonuria (PKU). With enough aspartame the blood level of a PKU-carrying pregnant woman could become high enough to arrest the development of her baby.[11] Dr. John Olney of Washington University in St. Louis reports that the aspartic acid in the product can penetrate the blood brain barrier and destroy brain cells, causing "silent" brain damage that may not show up for years and then appear slowly as obesity or neuroendocrine disturbances.[12] Methanol is widely regarded to be a poison. Dr. Woodrow Monte of Arizona State University reports that once in the body, it can undergo a chemical transformation and become formaldehyde and formic acid. Both are suspected of causing cancer and genetic mutation.[13]

Despite scientific evidence that calls the product into question, increasing numbers of complaints to the FDA, and the formation of citizens' groups like Aspartame Victims and Their Friends, aspartame is still used in more than ninety food products from diet soft drinks to sugarless chewing gum. It is sold as the tabletop sweetener Equal and called NutraSweet in diet drinks. It is added to meals in restaurants and "hidden" in countless packaged foods. Aspartame is a $1-billion-a-year business[14]—with millions of people innocently consuming it every day.

Thérèse is no longer one of them. She stopped her aspartame diet as soon as she found out about the chemical. Today she has returned to work and to her family responsibilities. Her seizures continue to occur, although they happen just a few times a year now, and according to neurological tests, she sustains permanent brain damage she believes was caused by aspartame.

Maria Carriaga's family has lived in their elegant adobe home near Albuquerque's Old Town for three generations, and ever since she was a child, Maria has shared the tomatoes, chiles, and squash from her three-block vegetable garden with her neighbors. Life used to be good in the Chicano barrio, but today the walls of the house that has stood for one hundred years are crumbling. The Carriagas can't speak in their backyard without shouting. The garden is barren. The fruit trees have withered, and in the last three months five of Maria's neighbors have died. Many people have respiratory problems, eye infections, and allergies. At fifty-three, Maria has been diagnosed as having liver deterioration and inadequate immune functioning. During one trip to the hospital, two doctors, independently of each other, exclaimed, "Do you live near a chemical waste dump?!"

Maria lives one hundred feet from Ponderosa Products, whose factory manufactures particle board. Every day, twenty-four hours a day, pipes spew dust soaked in formaldehyde into the neighborhood. Chemicals like benzene, toluene, and nitrate seep from unlined pits into the water table; noise from plant machinery is three times the accepted limit; and at night a distinct chemical odor envelops the entire community. Every month or so, what city officials term "an upset" occurs, an accident broadcasting black clouds of dust and chemicals all over homes, cars, yards, streets, and people for blocks around.

The problem began in 1974 when Ponderosa bought an orchard whose owner could no longer afford it and built a factory. To be sure, the area had accommodated lumber companies since the nineteenth century, and several, including Duke City, are still in existence, but through the years few problems had bothered the residents. Companies and community shared a respectful relationship. According to Maria, "You can call Duke City to complain, and they make an honest effort to change their behavior." But Ponderosa was different—in their reluctance to respond directly to the community, and in the toxic substances they used.

After the factory began operating, one by one the children in the neighborhood contracted allergies and asthma, including all twelve of the children who live on Maria's immediate block. Adults began to develop serious diseases of the respiratory tract, liver, and nervous system. Cars parked on the street were constantly covered with a half inch of sawdust, and no one spent time outside anymore. As Maria says, "We didn't understand what was happening. I live a hundred feet from Ponderosa, and *I* didn't know. I knew everyone was getting sick, and somewhere in the back of my mind I was slowly realizing it had to do with the sawmill. But I didn't know that the sawdust had chemicals in it, not until the church and the organizing project surveyed the neighborhood."

In 1986, El Buen Samaritano Methodist Church and the Southwest Organizing Project sent a questionnaire to the families in the neighborhood in an effort to identify problems needing attention. The survey revealed that the area's residents were concerned about sawdust blowing from the factory. Further investigation uncovered the fact that the sawdust contained formaldehyde and the plant was dumping chemical wastes into the ground. State and local agencies had tested the groundwater and found nitrogen levels in violation of water quality regulations. Nitrogen can cause cancer and blue baby syndrome. Traces of formaldehyde, ethyl benzene, methyl ethyl ketone, and toluene were also discovered. Although these chemicals are not classified as hazardous under federal regulations, New Mexico Environmental Improvement Division deputy director Richard Holland admits, "They do pose a risk to health."[15] Formaldehyde can cause irritation to the eyes, nose, and throat and is linked to cancer, neurological disorders, and immunologic deterioration.[16] Ethyl benzene and methyl ethyl ketone can cause leukemia, cancer, and damage to the liver, kidneys, and central nervous system. Toluene is the toxic compound in "sniffing glue." It is linked to fatigue, headaches, immunologic deterioration, and heart, liver, and kidney damage.[17]

Despite the fact that Maria is often exhausted, she is mentally sharp enough to understand the situation. "Now we may live in a poor neighborhood, and a lot of people here may not have a good education," she asserts. "But we're not *stupid!*" In planning production and disposal strategies, many corporations rely on a perceived link between poverty, poor educational opportunities, and diminished political clout to protest questionable practices. According to the Reverend Mac Legerton of Robeson County, North Carolina, "You take a poor rural county, add a high minority population . . . and you have the most vulnerable community for siting of massive waste treatment and disposal facilities."[18] A 1983 study conducted by the U.S. General Accounting Office confirms that hazardous materials tend to be used or dumped in areas with high minority populations and low income levels.[19]

Admitting responsibility, Ponderosa Products offered to buy Maria's house in 1985. Of all the houses in the neighborhood, hers was the closest to the plant, but the offer was not high enough to enable her to buy another house. According to Ponderosa director Ed Stewart, "[The neighbors] have good reason to come after us, but we have taken care of their problems. The last step [a bag plant] will remove once and for all any problems Ponderosa will have. Beyond that, it's just spite if we continue to have problems. We've done everything possible we can technically do."[20]

Maria's perspective is different. "The pain in my stomach gets worse as time goes on," she says. "I sleep all day and go to the doctor. It's sad. You know you have to die sometime. But it's sad you have to die because of a company, and you have to die slowly knowing you're being killed. Not only for yourself, but for your friends. We can't prove they're killing us. The doctors know but won't say anything. I just hope the plant gets closed down—before other people end up like me."

After finishing his residency in anesthetics at the University of California/San Francisco in the early 1960s, Carl Porter became a medical doctor. He later received a master's degree in public health and a doctorate in epidemiology. He was good. He became one of the first doctors in the United States to specialize in intensive care unit medicine, and in the operating room he successfully combined scientific training with intuition to know just what kind, how much, and how to administer anesthesia to his patients. In 1982 a job at a hospital in Redwood City was offered to Carl. He dearly wanted it. The future was looking bright.

But something was wrong. He had been tired. The arthritis that had started during his residency worsened, and his normal temperature dropped from 98.6 to 96 degrees. As time went on, Carl's health deteriorated further. He was dangerously sluggish on the job and often arrived late, holding up countless doctors, nurses, patients, and relatives. Occasionally he threw up during an operation. The hospital in Redwood City learned of Carl's reputation and withdrew the job offer, and Carl stood face to face with the fact that he was no longer a doctor, but now a patient.

Despite the opportunity he had to make use of the best of what the medical establishment had to offer, none of the specialists and internists he consulted had a clue as to what was wrong. Symptoms and test results did not fit neatly into accepted disease models, and too often the lack of diagnosis led doctors to pronounce that the problem was "all in his head." Meanwhile, Carl was pale and exhausted. He was too thin for his build, constantly nauseated and depressed, and he had repeated viral and bacterial infections that weakened him further.

So began what Carl calls "the hundred doctors": the frustrating and confusing quest for diagnosis, frustrating because doctors tended not to believe he was seriously ill and confusing because he did not know where to turn for help. He was on multiple antidepressants when his search took a turn. In 1983, a psychopharmacologist he knew had just heard of something called environmental illness: dysfunction of the immune system, subsequent deterioration of body systems, and susceptibility to secondary illnesses. The disease was brought on by overexposure to a specific contaminant or to many contaminants in one's environment. People got it, for instance, from being exposed to pesticides, working in chemical factories, or taking artificial hormones. For more complete diagnosis and treatment, Carl went to the Environmental Health Center in Dallas, Texas. Specialists there diagnosed him as having immunologic dysfunction and autoimmune disease. He now attributes his illness to twenty years of exposure to anesthetic drugs.

Since 1846 thousands of anesthetic compounds have been developed, compounds like ether, nitrous oxide, halothane, and pentothal. Their chemical structures differ, but they all serve the same purpose: to cause loss of consciousness. In the operating room these substances are used to render a person insensitive to pain and emotional distress and to support that person's life functioning during surgery.

It is only since the middle 1970s that the medical establishment has collected data chronicling the health hazards of anesthetic gases to the people working with them. Before this change in awareness, evidence was anecdotal. In 1920, researchers A. Hamilton and G. Minot reported gastrointestinal and central nervous system symptoms among World War I workers exposed to ether while making gunpowder.[21] A 1949 report tells of a surgeon, nurse, and anesthetist who contracted a variety of symptoms, including depression, headaches, anorexia, and brain abnormalities. When they stayed away from surgery for extended periods of time, they all recovered.[22]

Studies in the 1960s reveal increased mortality among anesthesiologists from malignancies of the stomach and lymph glands and from suicide,[23] as well as significantly more headaches, irritability, insomnia, and spontaneous abortions.[24] In 1974 a joint effort by the National Institute for Occupational Safety and Health (NIOSH) and the American Society of Anesthesiologists documented the following medical effects from occupational exposure to anesthetic gases: (1) significant increases in spontaneous abortions and congenital malformations in children of female doctors working in the operating room; (2) among offspring of male anesthesiologists, an increased incidence of liver disease and congenital malformations; (3) significantly more cases of liver disease and spontaneous abortions among wives of male operating room workers; and (4) an increase in kidney disease among female workers.[25]

Based on this study, NIOSH presented a set of recommendations to the medical profession, and by the early 1980s hospitals around the country had begun to change surgery techniques so that excess gases would be siphoned off, and precautions would be taken to avoid spills. In the United States, 214,000 workers benefit from these changes.[26]

Carl remembers when they were instituted. But for him, they came too late. By 1983 he had already been overexposed for twenty years. Today, despite drug treatments, specialized diets, vitamins, acupuncture, numerous orthodox and alternative therapies, plus ridding his home of possible contaminants, Carl's health is deteriorating. He is sensitive to all human-made chemicals, as well as to mold, pollen, dust, and animal dander. He has chronic fatigue immune dysfunction syndrome, hypoglycemia, and heart problems. His depression was so bad in 1985 that he took electroshock therapy and as a result now has selective amnesia. At forty-seven, Dr. Porter lives a life of isolation, peopled mainly by other doctors.

THE WOUNDS

An external influence necessitating
an abrupt change in adaptation
which the organism fails to meet.

—Abraham Kardiner, "Traumatic Neurosis"

DISCOVERY

All I did was go to work. Is that a crime?

—**Jesus Rives,** former tool mechanic
exposed to asbestos

For most of us the technological encounter is happening every moment of our lives. We live in what Marshall McLuhan calls "technological environment,"[1] and our very sense of existence is shaped by the continuous interaction. Think of it. Cars. Traffic lights. Freeways. Television. Telephones. Airports. Electric lights. Microwave. Librium. BHT. PVC. Chlordane. Polyester. Power lines. Pollution. Only recently have we begun to understand the effects of everyday technologies on our lives and health.

Most people made sick by a health-threatening technology are not aware at the time of exposure that the technology is dangerous. Because we do not exert choice over which technologies prevail over our lives or how, when the technological encounter is detrimental to health, it can occur in an atmosphere of ignorance. Because we are surrounded by a culture that assumes the benefits—or at the very least, the necessity—of technological progress, the event can occur in an atmosphere of innocence.

In 1949, June Casey was a sophomore at Whitman College in Washington State. At nineteen, she was bright, happy to be studying music, and eager to experience life. That December, nearby Hanford Nuclear Reservation, the nation's primary production center for weapons-grade plutonium, made an intentional, secret release of 5,500 curies of radioactive iodine—thousands of times more than the 15 curies released during the Three Mile Island nuclear accident. That Christmas, June went home to visit her parents in Oregon, and her "crowning glory," her head of thick, wavy brown hair, fell out and never grew back. She developed a case of hypothyroidism her

43

doctor described as "the most extreme case he had ever tested in his career." Despite chronic fatigue, she taught music, married, and did volunteer work for her local symphony and art museum. Later in life, June endured a miscarriage and a stillbirth. She never understood what had caused these disabilities.

Then in 1986, thirty-seven years after the event that changed her life, she stumbled across a small newspaper article revealing for the first time that a release of radiation had occurred at Hanford the very December her hair fell out and thyroid problems began. According to a 1950 monitoring report made public by the Department of Energy in 1986, the release was a "planned experiment."[2]

Similarly, in 1979 when Susan Hernandez went to work for the GTE Lenkurt electronics plant in Albuquerque, New Mexico, she was not told that the various varnishes, lamination compounds, glues, and cleaning fluids used on the assembly line would be detrimental to her health. Instead, for eight and a half years, Susan felt proud to be "working and making an important contribution to society." During this time she was exposed to dimethylamine, diethylene-triamine, diglycidyl ether of bisphenol A, lencast, and 4,4'-ispropylidenediphenol, epichlorohydrin resin. There were no face masks or protective clothing to shield her or her coworkers from the fumes and particles, and Susan worked at GTE for five years before the company even told workers to wear safety glasses. She only realized her work environment was dangerous when it became clear that, one by one, many of her fellow workers were falling sick with illnesses from allergies and dizziness to cancer and leukemia.

The crucial question that arises is one of knowledge. Who knows that a technology is dangerous? When do they know? How does a new technology get launched into public use? How complete are studies that research its potential impact? How influential? In some cases, as with the Dalkon Shield, the Pinto car, and leaking gasoline tanks, at the beginning no one really knows how safe or dangerous they are—not the inventors or the manufacturers, not the government or the consumers. No one has thought ahead to the possibility that they might have ill effects in the future, and sufficient testing and analysis have not been pursued. In cases like these, while neither purveyor nor user knows the dangers of a technology at first, eventually through unfortunate experience someone finds out. The discovery often pits defensive purveyor, who may not want to admit responsibility or invest in changing the technology, against wounded consumer, who may seek compensation for suffering or demand that the offending technology be banned.

In other cases, decision makers on the highest rungs of government, scientific, or corporate hierarchies do understand the dangers, but they determine that the "risk" to individual lives is worth the "benefit" to society, their own résumés, or their bank accounts. Seeing no advantage in confessing knowledge of the dangers, they often surround their technologies with secrecy. They tell neither workers nor public about potential problems, and as a result, people use dangerous technologies with no knowledge of risk.

The fact that asbestos could cause lung disease and death was known in the United States by 1918,[3] yet manufacturers persisted in employing workers in unsafe settings, avoiding responsibility through workers' compensation laws and legal corporate strategies. In the 1950s, Heather Maurer worked with her father cutting asbestos pipe for the family plumbing business. Her father died of multiple cancers, and her mother has pleural fibrosis today. "Believe me!" she asserts. "My father wouldn't have had his family work with the stuff if he knew it was killing us!"

Ultimately, we do not know the health effects of modern technologies because their developers and purveyors do not care to know. Our technologies are not created and chosen in an open, caring, or democratic manner, and we have not demanded that they be so. Rather their existence in the human community becomes, for both irresponsible developer and innocent consumer, *an unchosen fate.*

The discovery of the connection between a survivor's ill health and a technological event, then, occurs in an atmosphere of ignorance and innocence. Among the people interviewed in the study for this book, all but two expressed complete naïveté about the dangers of the technology they used or were exposed to. The listing on the labels of NutraSweet products informed Thérèse Khalsa that aspartame was merely "a sweetener," and Andy Hawkinson's officers encouraged the troops to enjoy their time on the island of Eniwetok.

The two people who felt endangered at all were only partially informed. They had enough knowledge to merit concern, but not enough to cause them to seek an alternative. George Milne was stationed on Bikini Island in 1946 to observe Operation Crossroads, atomic tests Able and Baker. As an amateur scientist with degrees in physics and chemistry, he had read as much as he could find about atomic energy, and he knew that the full health effects of radiation were unknown. Once on the islands, though, despite the fact that as beach master he was in a position of relative authority, George was never told how much exposure the troops received and there were no dosimeters to measure contamination. His uninformed ambivalence was reflected in island military policy. The lagoon where the men spent their leisure hours had two areas cordoned off: one marked "Uncontaminated," presumably safe for swimming and play; the other, just a rope's diameter away, marked "Contaminated."

Gay Ducey also felt caution. She knew that the herbicide she and her husband were applying to the church lawn in East Oakland was poisonous. She called the manufacturer to inquire about safety precautions, and just as they were advised, they showered thoroughly after applying the weed killer, put their clothes in plastic bags, and disposed of them at the city dump. There had been no warning about breathing the fumes, living on the grounds after application—or using the product while pregnant.

In an atmosphere fraught with ignorance and innocence about technological hazards, how do people discover the link between a technological event and their health problems? The answer depends, in part, upon the type of technology used and its health effects. Are effects immediate and obvious? Do they evolve through the years? Or do they appear, full-blown and unexpectedly, twenty years later? Making the connection between the event and illness also depends upon how much access one has to relevant information. Does one have books, reports, magazines, or knowledgeable people within reach? Are other people concerned and demanding to know? Or is one alone? Making the connection can also depend upon one's personality. Does one question prevailing cultural knowledge? Does one tend toward problem solving or intuitive discovery?

Despite the terrible pain of knowledge, it is agreed that the uncertainty of not knowing is worse. As noted by researchers K. Lang and G. Lang, "The worst kind of threat . . . [is] the dread of the unknown."[4] The most universal response of technology survivors is the desire to know.

For some people, the discovery is obvious and immediate. Diane Carter had a Dalkon Shield inserted in 1973 when she was living in Denver, Colorado. In 1975 when she first heard that the Shield could be dangerous, her doctor insisted she was doing fine and should leave it in place. Three years later, when she fell in weakness on the floor of her apartment, excruciating pain emanating from her womb, she knew the Shield was at fault.

For other people, discovery is a long and tedious process of piecing together experience, perceptions, and information. Without help from external sources, it becomes a personal journey. Betsy Berning is a survivor of both the Dalkon Shield and the overuse of antibiotics. She received the Shield in Cincinnati, Ohio, in 1971. Episodes of severe abdominal pain began immediately. For Betsy the process of discovery was one of learning to question the dictates of the medical profession and trust her own perceptions. One doctor repeatedly insisted that she was "imagining" the pain. He prescribed tranquilizers for six months to calm her nerves, but when he finally removed the Shield in 1975, to his surprise it had calcified and embedded in the wall of her uterus. She had severe pelvic inflammatory disease (PID). Seven years of antibiotics followed, a treatment that caused Betsy to get chronic vaginitis and the immunologic disorder systemic candidiasis. Two years later, her immune system severely weakened and, cervix ravaged by constant infection, she developed cervical cancer.

Although Betsy had acknowledged that the IUD was responsible for the PID in 1975, it wasn't until 1982 that she clearly saw the link between her total physical deterioration and medical technology. "I figured it out—or rather *it hit me,*" she explains, "when I stopped believing the doctors and started believing myself. And then I felt as if it had just happened!" Betsy's perception of the suddenness of the event is not uncommon. It took years for Maria Carriaga and her neighbors to manifest the health effects of the formaldehyde poisoning from the particle board factory, yet the moment they learned the cause, time compressed in their minds. They felt the contamination had *just* occurred. According to researchers Andrew Baum, Raymond Fleming, and Jerome Singer, because disastrous technological events are not anticipated, many people experience them as sudden, even when they have developed over time.[5]

Meryl Tavich discovered the link between migraine headaches and the birth control pills she took for ten years by pursuing her own medical research. She noted that her headaches began in 1965 when she first took the Pill. She observed their clear monthly pattern and located studies on the relationship between migraines and hormonal disruption. The various doctors she saw did not help her make the discovery. In fact, none of them knew or bothered to research the medical effects of oral contraceptives, and eventually Meryl came to know more about the subject than they did. She stopped taking the Pill in 1975, but the headaches continue to strike every month.

For some technology survivors, discovery of the connection between exposure and damage to health comes from external sources. Many people learn about it through the media. Atomic veteran Andy Hawkinson read in *People* magazine about the deaths of two military men who, like himself, had been at Eniwetok in the 1950s, and fellow veteran Gilberto Quintana, who was contaminated while on shore leave at Nagasaki, received a newsletter from the Disabled American Veterans describing the health effects of radiation from the bombings in Japan. Bookstore owner Pat Cody learned she was a DES mother in 1971 when she picked up the *San Francisco Chronicle* and saw an article entitled "Drug Passes Rare Cancer to Daughters."

Gay Ducey describes her discovery in poignant detail: "We didn't know why our son was born with birth defects or why Patrick and I had these chronic health problems. We occasionally tried to figure it out, but we didn't know. Then at Christmas in 1978, eight years after exposure, we were home visiting my family in Texas. Patrick picked up an article in *The New Yorker* by Thomas Whiteside. He'd written a book about Agent Orange called *Withering Rain.* As part of his continuing research on pesticide contamination, Whiteside had gone to a town in Italy called Seveso, which had been victimized by a cloud of Agent Orange—essentially dioxin. He returned to research and monitor the effects two years after the event. It was Christmas Eve, and my husband was reading late. He came upstairs at about two in the morning and awakened me, and he said, *'This is it. This is what we have.'* The health effects to the people and animals in Seveso were just like ours. I knew in an instant that he was right."

For some people, the connection between technology and health is made for them. A psychopharmacologist told Carl Porter about environmental illness when he learned about it from a medical resident. Perry Styles learned she was a DES daughter when a gynecologist detected abnormal cells growing in her vagina, a condition indicative of daughters of women for whom DES was prescribed, and indeed, Perry's mother later confirmed she had taken the drug during pregnancy.

Because the process of discovery can be haphazard and personal, rather than systematized and public, the question of accuracy arises. How can a person be certain that the technology used is responsible for negative health effects? The issue is complicated by the fact that we are all constantly being exposed to dangerous technologies. I mentioned that I am a survivor of medical technologies: artificial hormones and the Dalkon Shield. My health problems stem directly and historically from these two encounters, but various drugs have also been prescribed for me, including broad-spectrum antibiotics, at one point continually for five months, and the "morning-after" pill, which contains diethylstilbestrol. For the first twenty years of my life, my diet was spiked with pesticides and preservatives. I have lived downwind of nuclear testing and near high-voltage power lines, and I have breathed the pollutants of automobile and factory. I have slept on foam mattresses, gazed into television and computer screens, worn clothes made of petrochemicals, and had mercury amalgam fillings put in my teeth. I have lived in Cleveland, Ohio, where the air and water are toxic from industrial pollution; Fort Myers Beach, Florida, where trucks openly sprayed pesticides into the trees; San Francisco, a polluted urban area and one of the nation's centers of electromagnetic radiation emission; and New Mexico, where the nuclear industry spreads radiation via leakage wastes, uranium tailings, smokestack emissions, and unaccounted-for releases and accidents.

Despite the constant barrage of technological contamination we all endure, there are specific episodes that affect us more than others and can be identified. In a world of increasing numbers of health-threatening technologies, the personal experience of technology-induced illness arises as one of the most accurate routes to discovery. Few other routes exist. There is no holistic view of technology's health consequences and little systemic assistance from social agencies. Responsibility for discovery lies with the survivor, not with the purveyor of the offending technology or with society at large.

Unfortunately, extreme emphasis on scientific inquiry and technological development has robbed us of the validity of personal experience and subjective truth. Such modes are not "objective" or "rational" enough, and yet at this point they provide the most compelling voice for the sad predicament of modern technological society. Indeed, proof beyond a doubt that a particular technology has caused a person's illness is not always possible, but the fact remains:

Many people become ill after exposure to a technology—and they know it. Gay and Patrick Ducey know it. Thérèse Khalsa knows it. Betsy Berning knows it.

Response to the discovery that a technological encounter has caused deteriorating health is, of course, entirely subjective. The experience of becoming aware can be compared to the emotional process psychiatrist Elisabeth Kübler-Ross describes when people find out they have a terminal illness. In fact, many of technology's survivors do have terminal illnesses, or at the least, face incomparable loss.

Denial can be the first response. After Pat Cody learned that the DES that had been prescribed for her in 1955 could cause cancer in her daughter Martha, she applied, in her own words, "denial and repression." She tried to continue with her life as if no discovery had been made, comforting herself in moments of unwanted remembrance with the thought that the particular type of cancer described was supposed to be "rare." Eventually, though, the difficult news crept back into consciousness.

Pat's spiral of denial and awareness is common. Psychiatrist Mardi Horowitz points out the positive role that denial can play for people facing untoward events.[6] After the initial surprise, denial blocks registration of the new fact: "The cancer DES causes is rare." This numbing is then disturbed by the intrusion of the new information and feelings about it: "I did take DES. What could happen?" Then denial returns: "My daughter will be okay." The denial offers time to rest and assimilate. Soon the new information intrudes again: "The rare cancer they talk about was rare only *before* millions of women took DES." And the cycle repeats itself until the person, little by little, is able to face the full reality. This is a process of mastering loss. As Pat Cody says, "With something like that, it is so important, you can't bury it forever."

For some people, though, denial is the first and final response. They cannot allow the news to creep back in at all. In these cases, denial becomes not part of a process of acceptance, but an effective shield for life. Andy Hawkinson describes sitting at the bedside of fellow atomic veterans dying of cancer, leukemia, and heart disease, telling them their diseases were caused by radiation exposure, and being met with walls of silence and strings of dirty words.

Shock is also a common response to discovery. This can be the experience of the person who does indeed take in the news and believe it. It resembles "disaster syndrome,"[7] the complete lack of response initially experienced by survivors of tornadoes, fires, and other catastrophic events. For technology survivors, it is the sense of being stunned, dazed, and psychically removed from what has taken place.

June Casey describes the moment she read about the Hanford radiation release as "a knife in my heart," followed by shock. Love Canal resident Lois Gibbs describes her reaction as "stunned disbelief... the reporters talked about chemical seepage at Ninety-seventh Street," she reports, "then Ninety-ninth Street, and all of a sudden I realized that this was *my* side of the highway! This was *my* son's school! And I was just walking around in a daze. I didn't even talk about it with my husband. The whole thing was too much to comprehend. The reporters listed all these chemicals and their effects. There was benzene, which can cause central nervous system disorders, and lindane, which can cause miscarriages. And dioxin was there, and dioxin causes immune suppression. I was in shock!"

Then there is anger. "When we discovered the connection between dioxin and our son's disabilities," says Gay Ducey, "I was consumed, absolutely consumed, with the most killing, most cleansing rage I can possibly describe." There is anger at oneself, anger at the technology involved, anger at the institutions responsible for the technology, anger at the universe—as Saf-T-Coil IUD user Wendy Grace describes, "at society, at God, at everything and everyone—it wasn't any one thing."

Jesus Rives worked as a tool mechanic at Mare Island Naval Shipyard in Vallejo, California, from 1945 to 1979. As a result of a gall bladder operation in 1978, Sus underwent a complete medical checkup, including a new X ray to detect asbestos-caused diseases. He found out he had asbestosis. "I got very, very angry," he reports. "I got in my car and just took a long ride to get my thoughts together. Everybody has the same feeling, you know. I didn't think anything like that would ever happen to me. We see accidents. We don't worry about them. It's always someone else, but it hits home when all of a sudden—*boom*—they got you! I did nuclear work at the shipyard, especially in the latter years, and I was trained to guard myself from radiation. But nobody said one word about the asbestos. We were breathing it and the whole nine yards. It really got me mad to think

51

that I have it. The question that railed in my mind was: What is asbestos? I got something I don't understand, but I understand it's fatal. I was angry at everybody."

In the case of a community threatened by a pervasive technology, anger can turn into mass hysteria. Lois Gibbs describes one of the early reactions of her Love Canal neighbors when they learned about the toxic wastes beneath their homes and school. "We drove up Colvin," she tells, "then down Ninety-ninth. We went down Ninety-seventh, but I didn't see anything. Then at Wheatfield the road was blocked. Hundreds of people were in the street screaming, yelling, and talking . . . and burning papers in a bucket. I had never seen anything like it. They were like a mob."[8]

And there is fear. When Jane Woolf found out that the well water she had been drinking at her home in Michigan was contaminated by a gasoline leak across the highway, she also learned that the benzene in the gas could cause leukemia. "I was extremely worried," she says. "I had an overwhelming fear of death gnawing at me. Just thoughts running through my mind all the time. Who will I make my will out to? I wonder if I'll ever have any kids? I wonder if I'll ever have a family life? It was the idea that I could *die* from the exposure."

Another response to discovering the connection between health and a technological event can be relief. As June Casey describes, "It was the answer to the mystery of my whole life!" Although the news about his medical prospects was bad, Carl Porter cried, "Thank God!" At long last there was an explanation—and something concrete he could work on. To Gay Ducey the discovery was "terrible and wonderful" at once.

Gay also tells of the moment of discovery as a realization "of total violation, of technological rape." When Thérèse Khalsa learned about the aspartame in her diet foods, she felt "taken, abused, stupid, and dumb." Anika Jans remembers learning that the Dalkon Shield had caused, up to the point of discovery, two years of chronic fatigue and unrelenting fever. "I saw myself a vulnerable target for a cadre of omnipotent men wielding larger-than-life darts," she says, "and I knew that they had wounded me. I was no longer supported by normal reality. The earth fell away from my feet. I had become a victim."

LOSS OF HEALTH

Eventually I collapsed at work and had to be rushed to the emergency room.

—**Michael Wagner,** New Jersey homeowner
   exposed to formaldehyde in foam insulation[1]

All the Thule workers who were in the cleanup, they're all old men now. My husband is only 49, but he's an old man now.

—**Sally Markussen,** wife of cleanup crew member,
   1968 hydrogen bomb accident in Greenland[2]

Always, we notice the powder of the pesticides on the leaves. I was working in the fields from the beginning of my pregnancy. I was groggy from the Caesarean when the doctor told me, "I have bad news." He simply said Felipe was born without arms or legs. His father was crying. And my mother even more. When we first saw him we felt a great sadness. The doctors told me it was because of the pesticides.

—**Ramona Franco,** grape picker, Delano, California[3]

The first loss, and the one that precipitates all others, is bodily. Although the diseases caused by different technologies are different and the medical effects of a single technology can vary from person to person, what is universal is loss of health.

Atomic veteran Gilberto Quintana has endured a typical lineup of physical maladies he believes stem from exposure to radiation: prostate cancer, sterility, seizures, and bone deterioration. He de-

53

scribes his experience as "hell." Now over sixty years old, he has a pronounced limp, and on his face he displays the fatigue of a man overwrought by a lifetime of internal stress. "I was just thirty-eight when it all started," he says. "Since then I haven't lived one day without pain. I am always tired, always depressed. I do everything from the gut."

At the peak of her IUD-induced infections, Wendy Grace's physician said she was "as close to death as anybody I have seen." She remembers that time as one of helplessness and pain. After several surgeries, she could endure no more, and the doctor told her to lie in bed for nine months so the abscess festering in her belly could bleed through the skin.

After regulation termite spraying of her house with the chemical chlordane, Bliss Bruen suffered the miscarriage of "a stillborn little fetus," a child she and her husband wanted dearly and mourned for years.

Loran Calvert got asbestosis after working as a machinist and test specialist at a naval shipyard. Today his breathing is heavy and hoarse, and every day he worries that the disease will turn into lung cancer. "I do the best I can now," he says. "It could happen to me overnight."

The health effects caused by many technologies are still not understood. The slow march of laboratory studies and the quickly rising numbers of very sick people provide the material for research conducted by government agencies, industry, independent researchers, universities, and survivors themselves. As more technologies are proving dangerous, some causes and effects are emerging.

The principal biological effect of radiation exposure is the breakdown of cell structure. Oddly, radiation damage can be more serious at low levels than at high ones. Drinking contaminated milk or breathing background fallout can be more harmful than getting an X ray. High doses kill the body's cells, and for disease to develop, cells must be alive. Rarely killing a cell outright, low-level radiation scrambles cellular chemistry, rearranges genetic information, and leaves the cell vulnerable to the invasion of viruses, from common flus to more serious pathogens, like those associated with chronic fatigue immune dysfunction syndrome. Depending on the location of mutation, cellular disorganization can lead to rheumatic arthritis, leukemia, cancer, premature aging, sterility, premature births, congenital defects, cataracts, and death.[4]

54

Then there are chemicals and metals. At a 1971 meeting of the American Association for the Advancement of Science, a symposium attempted to rank nineteen major environmental stressors as to their effects on human health. Each technology was scored according to the persistence, range, and complexity of its threat. At the top of the list stood pesticides, followed by heavy metals.[5]

Exposure to pesticides and metals in water, food, and air can permanently disrupt nervous system functioning and cause brain damage.[6] Pesticide exposure is linked to hypertension,[7] Parkinson's disease,[8] epilepsy,[9] and cardiovascular disorders.[10] It can induce allergic sensitivities[11] and possibly liver disease.[12] The common apple-preserving chemical daminozide, if ingested over a lifetime, can cause cancer.[13] The pesticides captan, chlorothalonil, permethrin, acephate, parathion, dieldrin, methomyl, and folpet—detected by the Food and Drug Administration in apples, carrots, cauliflower, cherries, grapes, and peaches—are also carcinogenic.[14]

Dioxins, the herbicides known for their use in Vietnam, are also used in the United States on farms, national forests, urban parks, and lawns. They can be a potent immunosuppressant, leaving the body open to secondary infections, allergies, and autoimmune disease,[15] and a study reported in the *Journal of the American Medical Association* links use of dioxin by Kansas farmers with an eight-fold increase in non-Hodgkin's lymphoma.[16]

Industrial solvents are also extremely health-threatening. Chemicals like trichloroethylene, perchloroethylene, and 1,2-transdichloroethylene are found in electronics plants and dry cleaning factories, as well as in the water supplies of industrial communities like Woburn, Massachusetts. They can cause recurrent infections, immunologic deterioration, leukemia,[17] and neurological disorders that can result in panic attacks, personality imbalances, and spinal cord lesions.[18] Health studies of Woburn residents over a fifteen-year period after documented exposure to trichloroethylene, dichloroethylene, tetrachloroethylene, and trichloroethane show a twofold and threefold increase in birth defects and infant deaths.[19]

Exposure to the common air pollutants cadmium, lead, and carbon monoxide is linked to respiratory disorders, asthma, high blood pressure, and heart disease,[20] while contamination by many toxic chemicals can lead to a decrease in sperm density[21] or chromosome aberrations.[22] Exposure to benzene in gasoline products,[23] lead in air pollution,[24] PCBs (polychlorinated biphenyls) in electrical transformers and plastics,[25] and ozone from automobile exhaust[26] can result in the deterioration of one's immune response, as well as cancer, leukemia, and birth defects.

The flame retardant PBB (polybrominated biphenyl) can cause suppression of the all-important T-lymphocytes in the blood, weakening one's immune capability and increasing susceptibility to illness.[27] The new field of subliminal toxicology, which detects subtle changes in neurological function, reveals that people exposed to PCBs and dioxins can show memory loss, impaired mental acuity, and loss of coordination—all of which may be precursors to more serious disease.[28]

Formaldehyde is a gas used in insulation, particle board, plywood, furniture sealants, and carbonless carbon paper. It is inserted into carpets, cotton bedsheets, and deodorants and used in disinfectants and fumigants. According to recent studies, formaldehyde is linked to nasal cancer,[29] neurological disorders, respiratory disease, and immunologic deterioration.[30]

Asbestos has been used in pipe covering, brake linings, plaster, insulation, roofing, textiles, cement, paper, and felt. When the microscopically small fibers of asbestos enter a person's lungs, the body encases them in scar tissue, which, if pronounced, blocks the lungs from transferring oxygen into the bloodstream. The result is asbestosis. Other possibilities include mesothelioma (tumors on the membrane lining of the lungs) and cancer of the stomach, large intestine, kidney, larynx, and rectum. Exposure to asbestos can also affect the human immune response, leaving one open to secondary infections, allergies, and autoimmune disease.[31]

The overall effect of the introduction of vast quantities of chemicals and metals in the biosphere becomes evident when we compare cancer statistics. In 1900, cancer accounted for only 3 percent of the total deaths in the United States: that is, one in every thirty-three people. Since the introduction of thousands of new chemicals beginning in the 1940s, one in three people now contracts the disease, and according to the U.S. Toxic Substance Strategy, 80–90 percent of these may be induced by environmental contamination.[32]

Statistics on environmental illness also offer perspective. Environmental illness is the deterioration of the immune system to the point of susceptibility to allergies, viruses, bacterial infections, and imbalances of internal flora. By all accounts this condition has been on the rise in recent decades in the United States and other industrial countries. The Board of Environmental Studies and Toxicology of the National Research Council estimates that 15 percent of United States citizens are now hypersensitive to chemicals found in common household products.[33] Government research in West Germany indicates that one out of every four people in that country is sensitive to some food, chemical, or insect, and allergic reactions are appearing earlier in life than ever before.[34]

Adding to the dangers posed by chemicals and metals are other technological contaminants. Some researchers have found that exposure to the nonionizing radioactive pulses of video display terminals, electric blankets, and water bed heaters can lead to heart disease, male reproductive failure, miscarriage, birth defects, and depression.[35] Chronic exposure to the electromagnetic fields spawned by high-voltage power lines may result in disturbances in human biological cycles and diminished immune response—and a lessening of resistance against infectious diseases, leukemia, and cancers. Exposure to electroradiation can also cause birth defects and genetic diseases like Down's syndrome.[36]

Overdoses of antibiotics can disrupt the healthy balance of bacteria in and on the body, leaving a person open to infection by disease-carrying microbes or the overgrowth of internal flora like candida.[37] Oral contraceptives are thought to cause thromboembolic diseases like phlebitis, stroke, and heart attack; liver tumors;[38] hypertension;[39] and cancer.[40] New research suggests that artificial hormones, which include DES, steroids, and estrogen replacement therapy as well as birth control pills, can also contribute to the development of autoimmune diseases like lupus erythematosus, Graves' disease, and arthritis as well as lowered immune response.[41]

DES sons face an increased risk of infertility, structural abnormalities of the testes, and testicular cancer.[42] Daughters are two to four times as likely as nonexposed women to develop cervical carcinoma,[43] while some of them contract clear cell adenocarcinoma. Many DES daughters also sustain structural alterations of their reproductive organs, predisposing them to infertility, miscarriage, ectopic pregnancies, and premature births.[44] Mothers who took DES are 40 to 50 percent more likely to develop breast cancer than those who did not.[45]

Technologies added to what we eat and drink can also be health-threatening. Synthetic flavorings and food dyes constitute about 80 percent of all additives, and they have been linked to learning disabilities and hyperkinesis in children.[46] Animal studies show that the sweetener saccharin[47] can lead to cancer. The meat preservative nitrite can contribute to cancer,[48] as can the artificial hormones like DES, estradiol, progesterone, and testosterone that are used to raise farm animals in the United States.[49]

Vegetable gums like guar, carrageenin, and carob bean gum[50] and preservatives like sodium benzoate,[51] sulfur compounds,[52] and BHA (butylated hydroxyanisole) and BHT[53] (butylated hydroxytoluene) can cause allergic reactions—from hives and coughing to bleeding into the skin, migraine headaches, and mental illness. When molecules of the plastic packaging vinyl chloride chemically interact with foods, they can lead to cancer.[54] The chemical styrene, found in polystyrene-based disposable cups as well as in exhaled cigarette smoke and drinking water, can contribute to insomnia, nerve conduction abnormalities, chromosomal aberrations, and lymphatic and blood cancers.[55]

These are the known and suspected health effects of just a few modern technologies.

THE VICTIM

I feel as I'm sure a prisoner must feel who has been sentenced to prison for a crime he didn't commit.

—Survivor of Buffalo Creek mining flood[1]

A 39-year-old man died Wednesday morning after shooting himself while drinking beer and whiskey. Sheriff's Corporal said that the man took four bullets out of a .357-magnum revolver, spun the chamber, said "I'll show what's real," and fired the gun into his head. The friend told investigators the victim was depressed over an illness related to exposure to Agent Orange during the Vietnam War.

—Police report, quoted in *Albuquerque Journal,*
    December 17, 1987

Any person whose health is harmed by technology must grapple with the aspect of the experience called victimization. Under most circumstances, this person is living a normal life—making a home, working, serving in the armed forces, going to the doctor—when without clear warning or choice, she is thrust into physical suffering because of a technological event. A technological event is not an act of God, nature, or fate. It is not preordained. It is an avoidable, unnecessary occurrence wrought by human activity. This fact has a profound effect: It distinguishes the psychological terrain traveled by people who become ill for natural reasons from those who become ill because of human-made technology.

One metaphor technology survivors use to describe themselves is the canary in the mine. In the nineteenth century, mining bosses sent canaries into the mines to determine whether or not deadly vapors were present in the air. If the birds burst into song or flew back out, the air was adequate. If they fell to the floor of the shaft and died, the air was too polluted for people to enter.

Many technology survivors, particularly those suffering from environmental illness, perceive themselves as the "canaries" of technological society. They are the first to detect the poisons that will eventually affect everyone else. In fact, some technology survivors *are* "canaries," workers hired by various chemical corporations to detect leaking chemicals in and about their laboratories and factories. Like the birds of the last century, they sniff the air for traces of deadly chemicals like xylene, methyl isocyanate, and cyanide and in some instances, are even required to taste for leaks.[2]

Then there are technology survivors, like atomic veterans, who see themselves as "guinea pigs" used in experimentation. Their job has been to endure a technological event, be studied, and give feedback to the experimenters. Invariably, the feedback is negative. "Whenever you have technological development, you have risk," says Andy Hawkinson. "People invented the wheel, and then they got run over by it. The job is to figure out what's worthwhile and what's not. Nuclear weaponry may have brought a few benefits, but I'll tell you, my life as the guinea pig says loud and clear: It wasn't worth it!"

June Casey thinks of herself as the "sacrificial lamb." Irradiated by an undisclosed release from Hanford Nuclear Reservation, she feels she "was not even afforded the dignity of being studied for the effects." Instead, she was exposed and left in ignorance to deal with the consequences. Whether the analogy is canary, guinea pig, or sacrificial lamb—each represents a position of personal powerlessness in relation to people, institutions, and technologies that, altogether, work toward the destruction of human health.

As we know, most people are not aware of the dangers of the technologies that harm them. This lack of response springs, in part, from the fact that we live surrounded by a technological reality that has been presented to us as safe. In the last fifty years in particular, we have been encouraged at school and home, on the job and through the media to cherish the new technologies that science and engineering have produced. We have seen them enter our homes, neighborhoods, workplaces, government, and military—until now we exist in a world totally regulated by technology. The ability to perceive the dangers they pose is lost. Despite popular uprisings like the Luddites in England and labor movements in the United States and Europe, we have become so entrenched in our technological reverie that we assume the world has always been and will always be this way. As DES daughter Sarah Pirtle says, "We are like fish swimming in a polluted river. Because the filth is all around us, we don't have the full picture of how polluted the water really is."

The problem is not merely perceptual. It is practical as well. Even if we are aware of technology's dangers, we often have no choice. What other products can one buy but those sold in the stores? Where else can you get a job but in the local factory? Where else can a person live but near the nuclear power plant, in line of the electromagnetic radiation of the television station, or downwind of the paper pulping factory?

To make matters worse, the purveyors of today's technologies do not admit the potential dangers that they, in many cases, are aware of. Medical research before 1941 linked ingestion of synthetic estrogen to cancer in animals,[3] yet doctors and pharmaceuticals still developed and disseminated DES as a pregnancy drug. Then in 1971 studies linked it directly to human cancer,[4] yet today it is still used as a food additive in beef, for lactation suppression in women choosing not to breast-feed, and as the "morning after" pill. Despite the fact that they knew, the Niagara Falls Board of Education and the Hooker Chemical Corporation did not reveal to the families of Love Canal that they were living over a chemical waste dump. The United States government also knew that the radiation they subjected servicemen and civilians to in its testing program could be hazardous. They knew that leaks, releases, and accidents took place at nuclear facilities around the country, yet they did not tell those who would be exposed.

The result of such complex and all-pervasive forces is a woman who trusts her doctor when he says "This pill will make you feel better," a man who believes his workplace is safe, parents who assume their child's school is a place where she can work and play. The result is a growing population of people who—because they have no obvious reason to mistrust, no choice, or no information to tell them otherwise—are unknowingly harmed by dangerous technologies.

When a person is so harmed, he or she becomes a victim. The term is not popular. It interweaves the physical fact of violation with a person's psychological response to it, implying personal powerlessness. The term *survivor* is favored by many people who have undergone this violation because it does not avoid or conceal the physical fact of unjust harm done, yet it offers a range of options on the psychological level.[5] Whether one perceives oneself as a victim or a survivor, though, there are very real psychological developments one must face.

The sense of being "had" is one of these, and it is echoed again and again. "I'm mad. They took advantage of me," states Jose Luis Roybal, describing how the United States government ordered him to go to Hiroshima after the atomic blast. "Our health is determined by the whims of the doctors, the Food and Drug Administration, and the pharmaceuticals," cries Dalkon Shield user Sukey Fox. "I'm angry I wasn't given a choice," exclaims Susan Hernandez, whose immune system deteriorated after working with electronics chemicals. "How *dare* they do this to us! To them we are just numbers. We can be replaced. To them our lives are nothing!" The transition from healthy person to sick patient tends to pit one against the people immediately in charge of a technology's development or use. It enhances an us/them dichotomy and along with it, the feeling of being powerless, unheard, unseen, and angry.

One's assumptions about life can shatter. Suddenly the world is "out of whack," "upside down," "topsy-turvy," unpredictable. Whatever one has believed about stability, continuity, or propriety disintegrates as feelings of loss, violation, and uncertainty take center stage. "I felt confident and safe," reports Irene Baca. "I thought our government would never allow things like this to happen." "Life will never be normal again," says Carl Porter.

According to a number of researchers, the major component of this turnabout is the demise of one's belief in invulnerability—the it-can't-happen-to-me attitude.[6] Crossing over to an admission of vulnerability requires recognizing that the world is far more disorderly and unpredictable than one previously believed and that one may be helpless before the chaos. Marmika Paskiewicz became environmentally ill from exposure to chemicals on the job as a museum exhibition specialist. "The worst part for me," she explains, "was realizing that yes, this happened and yes, my job is making me sick. And I'm stuck in this toxic place. Despite all efforts to organize and change things, I'm getting no response! There's this attitude in the administration that's called 'being positive.' This means ignoring the problems. If I say I'm getting sick, *that's* 'being negative'! The worst part is feeling the victim, feeling helpless in an environment that is only getting more dangerous."

"You become vulnerable," says Lois Gibbs. "The worst part was watching the diseases come on and progress, watching the poverty happen, and feeling completely helpless to change them. Knowing every night you've tucked your child into bed, he's breathing polluted air—and there isn't a damn thing you can do about it."

One begins to see the world as a threatening place. After being rendered sterile by the Dalkon Shield, Diane Carter is so afraid of medical doctors that she refuses to see one ever again. Former sheet metal worker Nathan Robinson, recovering from lung cancer, gave himself a treat and bought a boat. But he is afraid to put it in the water. "I don't feel right catching striped bass on the bay anymore," he says. "I'm scared because of all the chemicals the fish are exposed to." A chronic attitude of hypervigilance develops, the sense that a world dominated by technology is not safe to inhabit. Anika Jans carries a protective face mask in her car in case of chemical spills on the road. Pesticide-poisoned attorney Robert McIntyre cannot live on his ranch in southern Texas because of regular sprayings at a nearby vineyard. Most of each year he lives like a refugee in his truck and when he senses pesticides in the wind, drives a hundred miles into the distance to escape.

Psychological studies indicate that survivors of rape live in fear of future rapes,[7] people who have been robbed are afraid they will be robbed again,[8] and cancer survivors live in terror of a recurrence of the illness.[9] Likewise, a majority of technology survivors express the fear that "it could happen again"—either from a similar technological event or, more likely, from one that the person does not understand and is ill prepared to guard against.

As appropriate as defensive feelings and behaviors may seem in a technologically threatening world, they also represent a psychological stance, a stance that can bring added stress to its bearer and, in its extremeness, even undermine attempts to find safety. In hypervigilance, says Irving Janis, "the decision maker . . . searches frantically for a way out of the dilemma, rapidly shifts back and forth between alternatives, and impulsively seizes upon a hastily contrived solution that seems to promise immediate relief. He or she overlooks the full range of consequences of his or her choice because of emotional excitement, repetitive thinking, and cognitive constriction"[10]—otherwise known as panic.

Diminished self-esteem can follow. Since psychological well-being is predicated, in part, on one's perception of stability and continuity, a person who feels violated and helpless can also feel worthless. "I'm not good enough," exclaims Dalkon Shield user Anika Jans.

The survivor becomes not just the victim of a technological event, but now the victim of his or her own psychology. Identifying with the impotence implied by the technological encounter, he feels worthless and in the isolation inherent to such negation, becomes deviant in a society that rewards conformity not just in machine parts, but in human expression. The question "Why me?" arises, "Why did *I* get exposed? Why did *I* get sick?" Since the offending event was human-induced and therefore avoidable, such questions arise frequently and with poignancy.

Answers too often relate to one's personal character flaws. "You blame yourself," says Susan Griffin. "You think there's something wrong with you for not being better in life. It's like an abused kid who blames herself because otherwise, why has she been treated so badly?" Harriet Beinfield reports that when her son was born with congenital heart defects and no thumbs, it took her a week to tell her parents. "I was so embarrassed and ashamed," she says. "I felt like I had failed, that I was inept for not being able to produce wholeness. I felt less than the least."

There can also be guilt, an ironic situation because it is the survivors of dangerous technologies rather than their perpetrators who feel it. Yet feel it they do. "I was bad," reports Sukey Fox after contracting pelvic inflammatory disease from the Dalkon Shield. "I felt like I was being punished for being bad and everybody was looking at me for doing bad things." "When my mother learned she had taken DES," Sarah Pirtle says, "she was ashamed. She felt tremendous guilt. Then when my son Ryan was born, she couldn't stand hearing the updates on whether or not he was surviving. She felt his problems were her fault. *What a travesty!* Here's my mother simply following the prescriptions the doctor gave her, and *she's* feeling guilty! The doctor isn't feeling guilty! The drug company isn't feeling guilty! When something terrible happens, it's easy to say 'How did I cause this?' It's like a spell cast over the victim."

The guilt stems from one's identification with responsibility, in some cases responsibility for having made a choice that led to exposure. "You made the decision to move here originally," explains Lois Gibbs. "You're still here, even though at this point you know it's dangerous. You could go live in the street. You could go to a church that takes in the homeless. There *are* some unpleasant alternatives. But you choose not to take them. It's a catch-twenty-two. Did I stay in Love Canal a year too long? What if Melissa gets leukemia? I'll always have to live with the guilt."

There is also responsibility for another person's well-being in the midst of catastrophe. Asbestos Victims of America executive Heather Maurer explains: "There's an embarrassment asbestos workers feel that they let themselves get exposed. There are feelings of guilt, like 'God, it's happening to me! I'm a fool.' And then they have to tell their families that they're dying, and that they endangered their wives and children because they brought the stuff home on their clothes. Some people don't like to talk about it because of the guilt."

In the face of the psychological demands that spring from the fact of victimization, suicide can loom as a viable approach. Former machinist Loran Calvert is a short, powerfully built man with a crew cut. He has asbestosis. He knows he will die and that his death will be excruciatingly painful. He considers "shooting myself in the head," yet holds back because if he dies by the disease rather than his own hand, his wife stands a chance of collecting damages from asbestos manufacturers. Carl Porter echoes this impulse. "I see no hope," he says. "In all honesty, I don't think I'm interested in being this alone and helpless for much longer." As the thirty-nine-year old Albuquerque man suffering from exposure to Agent Orange stated before he shot himself in the head, "I'll show what's real."

Loss of health, helplessness, hypervigilance, diminished self-esteem, guilt—if the labyrinth of psychological consequences stemming from the loss and victimization technology survivors experience constituted the only repercussion they deal with, it would be considerable. But the maze is not finished. Other aspects of the experience create even more demands for the psyche to deal with.

LOSS OF HELP

The most disturbing thing was not the illness itself, but the lack of appropriate response to it.

—**Susan Griffin,** prescribed birth control pills, cortisone drugs, and antibiotics in overdose

I went to my son's pediatrician, and I said, "Look, there are eight patients who have you as their doctor. All of them are under the age of twelve, all of them have a similar urinary disorder. Why is this? What do you make of the fact that you have eight patients who live within a few blocks of Love Canal who have *the same disease?!*" He said, "There is no connection."

—**Lois Gibbs,** Love Canal resident

Loss of health brings not only pain and suffering, it also brings loss of capacity and becomes a catalyst for a cascade of further losses of financial stability, social support, and a sense of meaning in life.

Losing one's ability to earn a living is a common result of technology-induced illness. As Dalkon Shield survivor Genevieve Hollander wryly states, "It's hard to make a living when you're not working." Putting it even more bluntly, a poster at the Asbestos Victims of America office reads: "Dying Is a Tough Way to Make a Living."

For some the loss is devastating. Because of bone deterioration and muscle collapse, atomic veteran Ricardo Candelaria was forced to give up his job as a soft drink distributor in 1970. He has been unable to work since. Until she died, he lived with his aging mother on *her* welfare check, and he spent his time dealing with health problems. After being contaminated by chemicals as a hydrologist in toxic waste management, Teddy Ostrow reports, "I lost my job, I

66

lost my house, and I had to sell all my possessions." When GTE assembly worker Irene Baca left her job because of complications from immunologic disregulation, she and her husband lost the dream of ever owning a home, faced medical bills they couldn't pay, and even sold the family furniture so they could buy food.

Many of the financial conflicts presented are irresolvable. Author Susan Griffin suffers from chronic fatigue immune dysfunction syndrome, a debilitating immunological disorder linked to a virus. She traces its onset to a series of technological stressors including birth control pills, the Dalkon Shield, antibiotics in overdose, cortisone drugs, and jet travel, all of which she feels diminished her immune capacity. She has been sick for over four years.

"When I did my taxes after the first year and added up what I could claim as medical," Susan explains, "it was eight thousand dollars. Plus I had to cancel my lecture tours, and I wasn't able to progress with the book I was writing. I can't even calculate. It must be something like thirty thousand dollars in the first year, at least. It's a vicious circle. There are all the costs, but at the same time you can't earn any money. The anxiety I went through had to be terrible for my health. I could never just rest and recover. I was always anxious about how I was going to get food and pay the mortgage."

In a story marked by a similar catch-22, Susan Hernandez tells about a fellow electronics worker, Wayne Lopez, whose exposure to chemicals caused his arms to turn crusty and weep pus from the elbows to the hands. Susan tried to convince Wayne to see the company nurse, but he was afraid that if he did, he would lose his job. "They would not relate his illness to the chemicals," she explains. "To them it's *you*. It's *your* body. It's not the job."

Then there are the hidden costs. Gay and Patrick Ducey's story exemplifies these. After insurance paid its share, they faced the enormous task of paying off 20 percent of nine operations for their son, plus ongoing medications for themselves. Gay also lost a year of work—and a year of income. On top of these obvious costs, the Duceys soon discovered that a child with a cleft palate is prone to ear infections. For the first five years of his life, Seth had medical bills every single month. Gay and Patrick also paid for speech therapy, schools and camps for children with learning disabilities, special sports equipment, and orthodontic work.

To add to the financial burdens that often afflict technology survivors, a person may also face loss of social support. Psychiatrist

Henry Vyner points out in his research on atomic veterans that this loss can lead to a transition from "being a connected social individual to being an isolated asocial individual,"[1] and as demonstrated in recent research, the presence of social networks, family systems, and support groups is a primary factor in a person's physical and mental well-being.[2] They provide a sense of belonging, bonds of intimacy, and feelings of mutual concern—landmarks that chart one's place in life. In his study of the effects of the Buffalo Creek mining disaster in West Virginia, sociologist Kai Erikson posits that the traumatic symptoms people experienced resulted from both the flood itself and the damage it wreaked upon their communal life, "from the shock of being ripped out of a meaningful community setting as well as the shock of meeting that cruel black water."[3]

Whether the technological event that harms one is a mass tragedy like the Buffalo Creek flood or a very private event like taking a prescription drug, technology survivors often lose the support of family, friends, and neighbors. In cases of community disasters, this loss stems from the complete disruption of social bonds and networks. In cases of individual technological exposures, it usually stems from psychological factors.

Connie Ruiz worked at the GTE plant in Albuquerque. When she became disabled with blackouts, memory loss, and slurred speech, her friends stopped calling—just when she needed them most. "They feel you are wrong," Connie says. "They say you are pretending."[4] Lois Gibbs's husband would not admit that their children's health problems stemmed from toxic chemical leakage, and when the federal government funded relocation to a motel, he refused to leave the house. Thérèse Khalsa's friends would not believe her when she told them aspartame caused her seizures. "They insisted that I was creating the disease," she says. "They wouldn't listen. They said I had a mental condition."

Loss of validation for the origins of an illness is not the end of the subject. Some people cannot deal with the gravity or implications of the situation. Living in a small town in the Southwest, atomic veteran Gilberto Quintana feels he must keep up a front of normality. "I can go around Pecos and tell people, but they don't know what I am talking about," he explains. "And they don't want to know."

Harriet Beinfield says that after her son Bear was born with a congenital heart defect and without thumbs she thinks may have resulted from pesticide exposure, friends outright abandoned her.

She now understands that the crisis, in particular the first five years, presented an extraordinary challenge. Harriet was living day to day in what she describes as "a traumatized state with a life-threatening predicament"—at a time when her peers were traveling, preparing for careers, and buying houses. "It was as if they saw me as a loser," she reports. "The worst part was feeling I needed support that wasn't there."

Technology survivors also endure a loss of validation from strangers. When Gilberto Quintana called the Los Alamos National Laboratory to research the relationship between his illnesses and radiation exposure, a scientist informed him with all the certainty of authority that there had been "no radiation at Nagasaki." When, for financial reasons, Susan Griffin dragged herself on a lecture tour of the Northwest, the coordinators of the tour treated Susan as though she were complaining rather than asking for help for a serious condition.

To Meryl Tavich, who suffers chronic migraine headaches catalyzed twenty-five years ago by birth control pills, the problem is caused by ignorance. "Because the connection between the disease and technology is not understood," she explains, "people tend to dismiss the illness, in my case by putting it in the category of 'women's complaints.' They roll their eyes back and think 'Oh, she's just being . . . you know.' If you say you're sorry you can't go, people think you don't want to go. Or they think you have an emotional problem. Otherwise intelligent people do not understand that what you have is *an actual physical disability.*"

Ignorance about technological dangers is not the only reason people withdraw from technology survivors or feel aggressive toward them. Such behavior also results from psychological projection. The unaffected person feels vulnerable and threatened by the fate of the survivor. To protect himself from his own feelings, he projects negativity onto that person, perhaps even going so far as to blame or punish the survivor. In the words of psychiatrist Henry Krystal, "The surviving victim becomes the subject of projection of a multitude of sins that we would rather forget and, therefore, much of the aggression tends to turn to the victim rather than the perpetrator."[5]

When Ricardo Candelaria gathered his courage to go on a live television talk show to discuss the legal and medical plight of atomic veterans, viewers called in and viciously accused him of being an unpatriotic traitor. Lois Gibbs recalls a man sitting next to her on a bus. He asked where she was going, and she responded she was returning home from a meeting at the White House. Curious, he asked what the meeting had been about. "Love Canal," she answered. The man shrunk into his seat, asking if she were contagious. Lois thought he was kidding. "Yes! I radiate chemicals!" she joked, and the man bolted to the other end of the bus.

Abandonment and rejection are not the only social losses technology survivors endure. If the offending technological event is a group experience, such as for workers at a factory or residents in a community, they can lose their friends to death. "What bothers me most is the daily deaths," says Heather Maurer of Asbestos Victims of America. "I was getting morbid the other day, and I thought if I had the funeral notices for everyone who died, I could paper my walls in a week. In a way I'd like to have them. Everyone who dies from asbestos disease, I want the funeral notice! The next time I go to see the state assemblyman, I'll just bring the stack, and I'll say, 'These are all the people who can't be here today . . . because they're dead.' "

Losing a leader can be particularly devastating. GTE worker Yolanda Lozano acted as chief organizer and "mother" to the nearly two hundred employees who banded together to sue the corporation. In 1987, Yolanda's allergies suddenly worsened to a debilitating degree, and she collapsed. She was diagnosed as having cancer and quickly died. Her death was a blow to the others. According to psychiatrist Robert Waelder, the presence of a leader provides an aura of protection and calm. Losing her can make group members feel abandoned and vulnerable.[6] As fellow worker Irene Baca says, "Yolanda was a jolly person. You could hear her laughter all the way across Department Three-twenty. It's scary to have her go. . . . Now I know this could happen to me."

Such a barrage of losses can be extremely stressful, but there can be even more to bear. Many technology survivors report a loss of validation from the people they look to with the most fear and hope: their medical caretakers. Technology survivors go to doctors for the same reasons most sick people do: to find out what is wrong with them, to learn the source of their illness, and to seek treatment. Loss of validation can originate in any one, or a combination, of these and results in what Henry Vyner refers to as the dysfunctional medical relationship:[7] a relationship between patient and doctor that does not serve the needs of the sick person.

Roberto Garcia's story provides a good example of the dynamic. Roberto worked as an electrician at the Mare Island Naval Shipyard in Vallejo, California. His job was to remove the asbestos covering the ships' pipes and then after repairing the pipes, to rewrap the pipes with fresh asbestos cloth. When his union sponsored a medical screening to diagnose asbestos-related diseases, Roberto was told he had asbestosis. Shocked, he took his X rays to his regular health plan physicians, but—he later learned—they had not undergone the special training necessary to detect the disease. Instead of admitting they didn't know, they challenged the diagnosis, insisting "You don't have asbestosis!" Roberto was left adrift and afraid. He had a diagnosis of a disease that was said to progress to death—but no doctor willing to help him with it.

Dysfunction can stem from a variety of sources. For one, unconscious prejudices can influence a doctor's behavior. Again, projection is involved. As a person harmed by technology, the patient becomes the recipient of doctors' unconscious projections about being "a loser" or "a victim." According to Krystal, ambivalence toward people damaged by disaster is frequently manifested as inconsistent and inappropriate administration of rehabilitation services.[8]

Sometimes the doctor blames the survivor by implying that she is a hypochondriac or his illness psychosomatic. When Diane Carter arrived at a Colorado emergency room with a 102-degree temperature, the doctor castigated her for "acting like a baby." Diane had severe pelvic inflammatory disease from her Dalkon Shield, and because of extensive scarring, she lost her ability to have a child. Carl Porter faced a similar reaction. Despite blood tests that showed he had the Epstein-Barr virus, nurses at the hospital greeted him with knowing nods and raised eyebrows. The most blatant example of this response was experienced by George Milne. The mental irregularities catalyzed by his brain tumor, plus his self-diagnosis that it was caused by exposure to radiation in the South Pacific, propelled doctors to dismiss him as "a psycho with an atom bomb complex." For technology survivors the experience is common, the underlying message clear: It's all in your head.

Technology survivors also encounter doctors who may know something about their condition but won't admit it, or may be able to perform tests but don't. These are usually physicians who have a personal investment in not revealing the source or nature of one's disability.

Irene Baca worked at the GTE Lenkurt plant from 1978 to 1980. Her job was to assemble transformers and lift heavy trays onto a fast-moving conveyor belt in a department reeking of chemicals. The stress of a minor car accident in 1980 exacerbated the immunologic disorders she was already experiencing, which ranged from arthritis and depression to chronic viral infections and allergic sensitivities. Her private physician recommended that Irene take time off from work to recover and that when she returned to work, she switch to a less stressful job. When she requested the time, GTE insisted that its own doctors look at her. After a cursory examination, they pronounced that nothing was wrong with Irene. Then, based on their report, GTE gave her an ultimatum: Come back to work now, or quit.

Lois Gibbs tells a similar story about a dermatology specialist at the Cleveland Clinic. He had worked with Vietnam veterans exposed to Agent Orange, and his expertise was skin diseases caused by dioxin. At first the doctor showed great enthusiasm about treating Love Canal residents, who had also been exposed to dioxin. Two days before the residents were to travel to Cleveland, though, he suddenly became evasive, telling Lois, "Don't send the people. They can't have dioxin-related problems." Aghast at this turnabout, she argued back, "How do you know?! You haven't seen them yet! You haven't taken specimen samples!" He flatly responded, "I'm telling you it won't be useful to you to send them. Period." According to Lois, when she spoke with other doctors at the Cleveland Clinic, she heard that the dermatologist had been pressured by New York State officials to drop the cases.

The rise of iatrogenic disease—illness caused by medical treatment—shifts the position of doctors from pawns, like the Cleveland dermatologist, in someone else's political game into willing participants avoiding responsibility. Fearing loss of credibility and malpractice suits, the medical establishment does not want to admit its practices can actually cause illness.

Genevieve Hollander's experience demonstrates the point. In the early 1970s, Genevieve was an active woman living in New York City and working in advertising. In 1973 her gynecologist on Long Island recommended she wear the Dalkon Shield. Not knowing its dangers, she consented. A year later she heard that the device could be harmful. The doctor told her not to worry. There was a slight possibility of infection, he indicated, but according to Genevieve,

"He never told me it could be a life-threatening infection, or one that I could never get rid of, or an infection that would cause me to lose every bodily part that has to do with reproduction."

In 1979 a series of episodes of pelvic inflammatory disease began. After a year of unsuccessful treatment with antibiotics, Genevieve had also developed a "grapefruit-sized" tumor on her ovary, and she underwent a total hysterectomy. In 1981 she initiated a legal suit against A. H. Robins, the manufacturer of the Dalkon Shield. As she amassed evidence for the case, she discovered that although the customary five years since she had last seen her doctor had not passed, *he had thrown away all her records.*

A final loss of medical validation stems from the inherent limitations of the Western medical system. Sometimes physicians invalidate a survivor's experience because their training has been incomplete. While many negative medical encounters may be traced to systemic causes, such as psychological posturing or the economic organization of modern medicine, this one stems from its philosophical base. The kind of health care most people in the United States have access to, allopathic medicine, is based on specific, but by no means all-encompassing, theories of health and disease. Moreover, this health care system enjoys a position of dominance over others such as homeopathy and naturopathy, not because it is the most complete or helpful, but because of past and current political maneuvering by its practitioners.[9]

The most basic tenet of allopathic medicine is evident in its name. The Greek word *allo* means "opposing"; *pathos,* "suffering." Allopathic medicine works by countering disease symptoms. It is based on what are called interventions, like chemical-based drugs and surgery, that single out a symptom and attempt to eradicate it. The approach does not delve into the original imbalances that produce symptoms in the first place. Seeing life from a mechanistic perspective and the body as a machine, it dissects the person into separate, supposedly unrelated organs and systems, emphasizes what is wrong with a single part of the body, and gives any negative development in that part a name.

As a result of this approach, allopathic doctors are trained to diagnose and treat only the categorized diseases that appear in their medical textbooks. When they encounter a set of symptoms that does not correlate with current medical knowledge, they face a problem. The problem is exacerbated by what Henry Vyner calls new diseases.[10] Just as new technologies invented in the last fifty years have had uncalculated and hitherto unknown social effects, so they have had uncalculated and hitherto unknown medical effects. These effects are not understood in nature or origin, and in their mysteriousness, doctors rarely validate them.

Lois Gibbs's son went to school in a building located directly above the Love Canal toxic dump. He developed what appeared to be asthma, epilepsy, and a urinary tract disorder. The doctor diagnosed these as three separate, unrelated conditions with no hint of a common origin, either in his body or in the environment. "But Michael's epilepsy wasn't really epilepsy in the traditional sense of that term," explains Lois. "It was a seizure disorder due to toxic exposure. But the pediatrician didn't know anything about toxics or about the interconnection of his supposedly unrelated illnesses." Susan Griffin echoes this frustration. "I feel trapped by the system. If my doctor has a name for something and she has a drug for it in her desk manual, that's fine. But if she can't name it, she acts as if my suffering isn't real."

Based on rigid categories of symptoms, allopathic medicine brings little subtlety to its understanding of the disease process. As Susan says, "I kept saying, 'There is something else going on in my body that you aren't addressing.' I *knew* it! But there was nothing she could do until the symptoms got so bad that I had several major breakdowns that corresponded with categories in her desk manual. That's the way Western medicine is. It can't detect imbalance until you are flat-out ill, and it can't validate it unless it's written up in the book."

The practice of allopathic medicine can also alienate the doctor from the sick person by placing the highest premium on the doctor's authority and invalidating the sick person's knowledge of the healing process. Interventions are extreme and dramatic, sometimes going so far as to replace the body's own functioning or its capacity to heal itself. Also, using technologies like radiation therapy or dialysis, which are beyond the layperson's ability to comprehend, interventions must be ordered and supervised by "the expert."

One such expert was one of Susan Griffin's doctors who, despite lengthy discussions about Susan's history with ulcers, prescribed a drug contraindicated for "gastrointestinal distress." Or Betsy Berning's doctor who gave her seven years of antibiotic treatment for an illness exacerbated by antibiotics. Or the doctors Benina Berger

Gould saw who prescribed oral contraceptives, IUDs, artificial hormones, a complete hysterectomy, estrogen replacement therapy, and mammograms—one technology upon another, each addressing the problems created by the technology used before it.

Just as technology survivors can lose validation from health care providers, they can also lose it from the institutions whose purpose it is to serve them, institutions like state and city government, the military, public hospitals, and mental health agencies. This is, to paraphrase Vyner's term, the dysfunctional service organization relationship, and it reflects Henry Krystal's insights about inadequate rehabilitation to damaged people.

Many former servicemen and servicewomen exposed to war technologies like Agent Orange, nuclear testing, and biological warfare assert that the United States Department of Veterans Affairs (formerly the Veterans Administration) serves them poorly. The avowed mission of the DVA is to "provide compassion, quality care, support and recognition" to the 27.7 million men and women who served the country and their 78.4 million dependents.[11] One way it attempts to accomplish this is by assuring medical care and psychological counseling in its hundreds of hospitals around the nation.

Despite constant pain, fatigue, and lack of transportation, Ricardo Candelaria has gone to his local DVA hospital dozens of times to receive these services, yet he never has. Ricardo's first frustration is that the department will not admit that his disabilities stem from exposure to radiation. Once, he says, a DVA doctor took him aside and told him that in all probability radiation caused his sterility, skin lesions, bloating, fatigue, and constant pain. Professionally, however, the doctor would not admit the connection. As a result, any help the DVA might offer Ricardo falls under the category of nonservice-connected disability. Such an evasion is not only an assault on Ricardo's sense of justice: It leaves him with more of the bill to pay, a titanic feat for a man who hasn't worked in over nineteen years.

Similarly, veteran Jose Luis Roybal was given a 10-percent rating of service-connected disability, which means his health problems are officially measured as 10 percent caused by events during his stint in the armed forces. He claims that if health can be numerically rated, he deserves a 60-percent rating or better. He has been completely disabled with radiation-related illnesses ever since returning from postwar duty in Hiroshima. Recently, the DVA denied any portion of his claim for a tumor growing in his neck. Without treatment, it continues to grow.

According to a 1984 survey of atomic veterans, 99 percent are convinced that the DVA has not given adequate medical care or compensation for radiation exposure.[12] In 1987, 64 percent of all radiation claims filed were denied,[13] and the government continues to deny the connection between many illnesses and exposure. The Pentagon asserts that 99 percent of atomic veterans received less than five rems of radiation, and a 1985 National Academy of Sciences study claims that atomic veterans do not have higher cancer rates than other people.[14] Indeed, the facts that the United States government did not uniformly provide measurement of radiation at tests or that measurement technologies were not highly developed in the 1940s and 1950s have made a cause-effect link between exposure and illness difficult to verify. In its own interest, though, the DVA has steadfastly denied medical research that establishes a link between low-level exposure and numerous health effects.[15] Yet the passage of the Radiation-Exposed Veterans' Compensation Act of 1988 designating that at least thirteen cancers are radiation-related reveals that the evidence can no longer be ignored.

Ricardo Candelaria's second frustration with the DVA concerns its chronic lack of responsiveness to his case. Such a dysfunctional relationship between veterans and the organization created to serve them is common. According to veteran wife Judy Beck, featured on a 1987 *20/20* television exposé, "You go to the Veterans Administration, and you don't know *that's* your enemy."[16]

A rumor circulating around the waiting room in the Albuquerque office says: "They want you to get fed up and stop trying." It elucidates Ricardo's experience. He originally made an appointment with a caseworker to apply for nonservice-connected disability. The caseworker did not show up at the appointed hour and place. Ricardo remade the appointment five more times over a four-month period, and on each day the caseworker did not arrive. Finally, the receptionist explained that Ricardo's caseworker had not worked in the department for at least three months.

Ricardo started again. He made an appointment with a new caseworker, and she did not show up. A second time she explained that she was too busy to see him; a third time she left work early but did not bother to reschedule. Trying to get medical attention became Ricardo's main activity in life. Finally, when he became bedridden with bloating so severe the skin on his legs cracked open, the effort proved too much.

Other service institutions respond to the needs of technology survivors with similar disregard. Sometimes their responses can be irrelevant and crude. Teresa Juarez of the Southwest Organizing Project describes an official from the Alburquerque Environment Health Department visiting a home near the Ponderosa Products particle board factory: "Look at all the sawdust in here!" exclaimed the homeowner, pointing to the dust broadcast all over her yard and car and migrating into the house through cracks in windows and doors. "That's not sawdust," the official pronounced. "That's your housecleaning."

At other times government responses can be evasive. In 1982 the tap water in Charlotte Mock's district of Albuquerque turned oily and metallic tasting. Charlotte and her neighbors found little black flecks floating in the dishwater and toilet bowls and noted that "whatever is in that water, it eats the chrome off the bathroom fixtures." With research they discovered that this same predicament was affecting some eighteen thousand local residents of the Westgate Heights area. At first the city Water Resources Department claimed the problem was due to a tarlike residue left over from cleaning the edges of the reservoir. The state Environmental Improvement Division said it could be mineral oil used to lubricate the pump shafts. But if the problem related to either of these, it would have subsided in time. It did not.

Charlotte and her neighbors called a meeting of Westgate residents. They collected samples of water from a number of homes and sent these to the New Mexico Scientific Laboratory Division for analysis. Much to their surprise, every time they called for results, officials claimed that the samples had been lost. They would send more, and these in turn would disappear. Charlotte also noticed that whereas the local reservoir had previously been open to the public, now the fence surrounding it was bolted shut with a padlock.

Alarmed, Charlotte sent water samples to the Centers for Disease Control in Atlanta. They promptly reported a significant presence of contaminants in the water, including the carcinogens xylene, benzene, and trihalomethene. In the meantime, the state Scientific Laboratory Division got back to the neighborhood association saying the water was fine, although the tarlike particles could not be analyzed "by the methods at our disposal."[17]

Charlotte's group also tried to alert city and state elected officials. After much coaxing, the mayor spent two days at the Water Resources Department. He came away with official reassurances and no more information than the neighbors had garnered. Charlotte tried to run for public office to bring the problem into open discussion, but she lost. For her own peace of mind, she requested that the city attach a filter to her home's water line. She hauled filtered water for several months and finally purchased a heavy-duty water filter.

In the last few years, the water has improved somewhat, at least to the naked eye, but the experience as a whole leaves her feeling invalidated by the very agencies she had perceived as serving her. Today she asks some hard questions. "Does the problem originate with abandoned underground gasoline tanks? Is it from toxic wastes buried around Kirtland Air Force Base? Or is it a natural occurrence from hydrogen in the rock around here? And why is there such a high incidence of leukemia in Albuquerque? We don't know the answers—*and we should!* Why did the state lab say the water was okay when it obviously wasn't? Why did they keep losing our samples? Why did they lock up the reservoir? And why did the politicos refuse to get involved? I am deeply afraid they just want us to go away—like at Love Canal."

At a time when social validation and support are needed more than ever, they are too often lost—from friends, family, employees, coworkers, the medical profession, and service organizations—and life, already made difficult by illness, becomes harder to manage and bear.

LOSS OF HEROISM

I can't do anything for anybody. I can't even do anything for myself. What purpose do I have in this life?

—**Ricardo Candelaria,** atomic veteran

A person's sense of meaning in life is central to psychological health. The cultural anthropologist Ernest Becker calls the significance we infuse into life our hero system. This consists, first, of a belief system and, second, of a means to enact that belief system. To Becker it is immaterial whether one's hero system guides one to perform bold acts or to earn a living, to live according to "a universal plan" or to work in politics. What is important is that it gives a person a sense of meaning: a feeling of value, of usefulness, of connection to forces greater than one's individual self—of heroism.[1]

Most technology survivors lose all or part of their hero system. Long-standing, fundamental beliefs about themselves and the world can shatter into irretrievable fragments. One's identity can be the first to go. Psychologist Benina Berger Gould experienced a loss of sexual identity when she underwent an emergency hysterectomy. After having a family, Benina was advised to take birth control pills and later to wear the Dalkon Shield and Copper-7 IUDs. She developed severe pelvic pains, which were diagnosed as endometriosis (and were later, in surgery, found not to be). To treat the condition, a doctor prescribed massive artificial hormones. She then developed what was diagnosed as a cyst, which, the doctor asserted, would explode in her abdomen if she did not have a hysterectomy within ten days.

She did. Her response to the surgery was a feeling of piercing loss. "I would never again see that part of me that was female," she recalls. "There was a hole in my belly. No more cervix. No more bleeding. I missed the bleeding, as if my blood every month gave me a rite of passage into being a woman. My mother was afraid I would start aging. I was afraid I wasn't a woman anymore."

Loss of other crucial aspects of one's sense of self can also occur, for instance that of being capable and independent. Sukey Fox had the Dalkon Shield prescribed for her in 1972. Despite the IUD, she became pregnant. Sukey had an abortion, but during surgery the doctor was unable to locate the device in her womb. He assumed that despite the fact that Sukey never noticed, the IUD had been expelled from her body. In 1980 she developed pelvic inflammatory disease with constant fevers, excruciating pain, hot flashes, and tubes too scarred for future conception. Doctors at the Mayo Clinic in Rochester, Minnesota, examined Sukey and concluded that the disease was caused by the IUD. Either the inflammation had festered at a subclinical level for years and then erupted in 1980—or the IUD that caused it was still in her body.

Today Sukey still experiences pain, and she has hot flashes she feels signal early menopause. Apart from the physical pain, fear, and loss of work she has endured as a result of her technological encounter, Sukey reports that the experience challenges one of the most cherished aspects of her identity. "My image of myself is that I'm a healthy person," she says. "I exercise regularly. When I fill out job applications I always write 'Excellent health.' But in fact I've had all these medical problems all these years."

Similarly, Laura Martin-Buhler thinks of herself as a giving person. She is an active member of her church and a dedicated mother. As a child in the 1950s, she lived in various towns in southwestern Utah. At that time the government was planning its aboveground nuclear testing so that the fallout would waft in the Utah wind, landing as pink dust on cars and sidewalks. In 1960, when she was twelve years old, Laura developed a precancerous condition of the thyroid. A surgeon removed all but a sliver of the organ, and Laura began a lifetime of taking thyroid medication.

The thyroid secretes a hormone into the bloodstream that regulates metabolism and influences emotional balance. Laura describes "life after you've had your thyroid removed as an emotional roller coaster." If she doesn't take enough medication, she becomes too exhausted to care about other people. If she takes the recommended dose, she becomes too angry and picky—not at all the giving person she enjoys being.

For some technology survivors the loss that brings the most pain may seem minor. After forty years June Casey still feels humiliated that her hair fell out from radiation exposure. To this day she describes the thick, wavy hair she no longer has as her "crowning glory." Compared to her suffering from hypothyroidism, various miscarriages, and the threat of cancer, this problem may appear relatively insignificant. On a daily basis, wearing a wig does present inconveniences and moments of embarrassment, but the real source of pain is symbolic. For June beautiful hair represents her identity as a woman and lies at the heart of her meaning in life as wife and mother.

Many technology survivors lose their ability to be what they want to be or are expected to be. They can no longer perform well or they cannot perform at all. They lose their work, and they lose their identity in the world. Carl Porter lost his job as an anesthesiologist and became an invalid. After returning home from Hiroshima, Jose Luis Roybal got married but could not provide for his new family. When he was poisoned by pesticides, Danny Morris gave up a lifelong dream of attending Harvard Law School. His immune system weakened to the point of sensitivity to all synthetic materials and many natural ones, he began an exhaustive search for an uncontaminated place to live. When he found one in Texas, he began a life of living in a tent and spending all his time gathering food and water for survival.

Another devastating loss that threatens a person's sense of meaning is the ability to have children. Many technology survivors lose their biological ability to reproduce, endanger their lives if they do, or endanger the life of the child if they do. Causes of this loss range from use of an intrauterine device and DES to exposure to chemicals and radiation.

Ricardo Candelaria's experience tells the story. Ricardo grew up in the 1930s in the Hispanic barrio of Albuquerque, New Mexico. It was a slow-moving time when people farmed the land, ground corn to make tortillas, and sold chiles on the street. The core of the tight-knit community was the family, and Ricardo looked forward to having one of his own.

In 1950 he was proud to be accepted into the United States Marine Corps. This was an opportunity highly valued in the barrio. It was perceived as heroic. Shortly after boot camp Ricardo was sent to Yucca Flats, Nevada, where he was ordered to witness the Ranger atomic test from a foxhole a few miles away. He describes the experience: "You were supposed to drop down into the hole, dig yourself in, and put your gas mask on. They told us when the count reaches zero, the flash would come first—bright blue—and if we wanted to see it, open one eye. Well, I closed both eyes as tight as I could, but I still saw that blue light right through my eyelids. They also told us if we wanted to see the blast, count to twenty before getting up. To be sure, I counted to thirty. Then I stood up, and I saw the mushroom going up, the bright red streamers of fire flying down from it, and a doughnut of fire billowing out in all directions. That doughnut was so big, I can't tell you, and it was coming toward us fast. You could see the heat rising up and the dust forming into a mushroom pattern in the sky.

"My mouth fell open. I didn't know what was happening. The helmet blew off. You were supposed to lean up against the wall of the hole for when the blast came, but I completely forgot. I was just looking at that cloud. Then the blast came and threw me against the hole. I'll never forget it."

After waiting in their foxholes for a few hours, the soldiers were ordered to approach what Ricardo appropriately calls "dead zero." Then the military technicians detected amounts of radiation that were, even in an experiment whose purpose was to irradiate people, too high. The desert soil was black and cracked. Military equipment placed near the blast site had powdered into the ground. At one hundred yards the troops were ordered to retreat.

Indeed, Ricardo has never forgotten this event. Not only is the image of the rising cloud emblazoned on his mind, but in 1956 he discovered that he was sterile. His fiancée left him, and he has not had an intimate relationship since. Today he explains with hurt in his voice: "If I'd have known about the radiation, I wouldn't have gotten into that trench. I'd have been court-martialed and gone through whatever trial and punishment they had. If I had to come out of the service with a bad discharge, then I'd have come out healthy—and I'd have been a father. And I'd have been a *good* father." His shoulders shaking in sorrow, he cries: "I wanted a family so bad. You don't know how much I wanted a family!"

Gay Ducey's pain is also for the children unborn. "One of the worst parts of this experience is a personal feeling of loss," she explains. "My husband and I had these two marvelous, inspiring children. But we had planned to have two more. I miss them. I feel their presence, those other two. As technology survivors, you have to make decisions that other people don't have to make."

Andy Hawkinson's confusion as a father centers on the children he did have. "I don't know if the root of the problem can ever be reached," he says. "The pain, the suffering, and my perpetuation of that by fathering children! I did so in 1960, '62, '64, and '65. I was exposed to radiation in 1957, but I didn't know about it until 1977. My daughter just got married, and I wonder: How can I be a good father? Should I tell her she could have a deformed baby? Should I *not* tell her?"

On top of these many losses to one's sense of heroism, faith in once-trusted institutions can also be lost. This experience is different from the loss of support many technology survivors experience from institutions they thought would help them. This is the loss of a sense of connection to, respect for, or loyalty to the institutions survivors perceive as directly responsible for hurting them. The breach of trust that results can present one of the most profound losses a technology survivor has to confront.

Genevieve Hollander grew up believing in the authority of the medical establishment. When she agreed to use the Dalkon Shield, she did so with trust for that authority. Her belief was shattered not only by the fact that the IUD caused pelvic inflammatory disease so severe she had to have a hysterectomy, but also by the deceit and disrespect she encountered from her doctor when he threw away her records. Genevieve's response to these assaults on her beliefs is social cynicism. "The thing I've learned," she proclaims, "is that there is no justice!" She is transported from a sense of trust into a state of outrage.

Unless pressed to discuss it, Andy Hawkinson would be the last person to reveal his anguish about the breach of trust he feels. It is too painful. When he went into the Army in 1957, he was a "red-blooded patriotic American boy." In 1967, Andy volunteered for a second tour of duty so that he could serve in Vietnam. He felt his country needed him.

Andy describes the loss he experienced. "I was in the Army thirteen, maybe fourteen years. The Statue of Liberty. Red, White, and Blue. My Country. I have often asked myself if I had to lose something, in what order of progression would I be willing to give it up? Well, I would probably sacrifice my mother first . . . my wife second . . . my children third—before I would ever, *ever* sacrifice my country. But then not to experience the same reverse loyalty from my government when I'm willing to give it so much—this is a rejection I'm not prepared to handle."

It is a rejection that has propelled Andy to become a veterans' activist and join liberal political organizations he never would have supported before. Despite his transformation in worldview, Andy has some compassion for the government. He believes he understands the rationale behind the nuclear weapons testing program. "I know why we had troops in those tests," he explains. "It was combat indoctrination. It's why we put troops through a gas chamber or expose them to live fire. It's why we make them go out on a twenty-five-mile march or take apart a rifle. These are things that teach a military man to function. Back in the 1950s the military wanted to know what would happen to a man when he was exposed to a nuclear detonation. War was envisioned as nuclear, and the big question was: Can a soldier pick up and go on fighting after the blast?"

Andy's perception of why testing took place does not, however, bring him forgiveness or peace. To this day one lone question plagues him. "Okay," he says, "so after a test or two they found out the answer to the question. *Then why the hell did they keep doing it again and again and again?!*"

What were once rewarding relationships between Genevieve Hollander and the medical establishment or between Andy Hawkinson and the United States government become hurtful relationships induced by a technological event and corroded by differing beliefs about that event. Because of the new relationship, which Genevieve calls "a bad divorce," she, Andy, and other technology survivors are catapulted into foreign psychic territory. This is a state of mind characterized by the shattering of a formerly viable hero system and fraught with bewilderment about meaning. It is a sense of being both outraged and lost. The experience is reported by survivors of technological disasters ranging from the Buffalo Creek flood and the Hyatt Regency skywalk collapse to Love Canal.[2] Says pesticide-poisoned Bliss Bruen with pain quivering in her voice, "I've learned absolutely not to trust any government agency. If they say it's safe, you can believe it's not. And definitely don't trust anyone who's selling a product."

Frank Kaler concurs. For five years he struggled unsuccessfully to get government help to close a landfill that was contaminating his family's drinking water with lead, benzene, carbon tetrachloride, and radioactive waste. "I used to take my children to see the parades," he says, "and I told them always take your hat off, stand at attention, this is your flag. I believed there was justice. I was naive. Today I teach my children that you can buy your share of American justice *if* you have enough money to pay for it—and no other way. The laws are made by the rich and powerful to protect the rich and powerful, and the richest and most powerful are the industrialists from large corporations. *They . . . run . . . this . . . country.* "[3]

Loss catalyzes bereavement, and whether for the loss of a meaningful job, beloved friends, one's sense of self, or belief in one's country, bereavement is an emotional process that is all-consuming. Its core is separation. The bereaved is separated from the familiar bonds that previously defined existence. One is thrust into an unknown reality where needs are not met in obvious ways and meaning is not readily apparent. At face value, loss can catalyze anxiety and a state of helplessness, but there is more to consider. The separation can also penetrate into deeper chambers of one's being, reawakening memories of separation during the first years of life when, indeed, an infant could not survive if separated from other people. When a negative technological event occurs, one's immediate physical survival may in fact be threatened, and it may not. Regardless, the losses occurring in the technology survivor's life can touch off primal fears and the belief that—right here and right now—as Anika Jans describes, "it feels like life is over and all is lost."

One's emotional response can be intense. There can be pangs of longing, panic, anxiety, and helplessness. There can be anger and regret or an incessant welling of sorrow. In the context of such all-consuming emotions, denial can be a useful, even necessary part of the process of mourning. It can provide the bereaved with the wherewithal to face life anew within the limitations of loss.

But not everyone is capable of undergoing successful mourning. For some technology survivors the losses they must endure may be too overwhelming. Andy Hawkinson candidly describes his use of denial. "In coping with it all," he says, "I live a great big lie. And that lie is made up of a whole bunch of little lies. I keep telling myself, 'This isn't a problem,' 'That isn't a problem.' I lie to myself saying, 'Sure, you can handle it. You can overcome it. It's not going to wear you down.' And periodically, whether it's biological or emotional, something ruptures and I have to live through another trauma. Then I get back on my merry-go-round of lying until the next episode.

Interspersed between them I do some things that are constructive. I never become destructive because along the way I find that I can tell myself a lie to answer a lie which gives me the strength to keep fighting."

When survivors remain in denial, they risk falling into psychological patterns that express their grief in unconscious and often destructive ways. One pattern that can follow incomplete resolution of bereavement is psychosomatic illness. Another is psychological disorder, from chronic anxiety and depression to use of drugs and alcohol. A third destructive pattern may be expressed through a survivor's personal relationships by excessive neediness, bonds that are infused with denial, or social withdrawal.

The result is hard times. No matter how great or small their losses, technology survivors are catapulted into a state of bereavement that can be exhaustive, last for years, and leave one bereft of resources. As Andy asks in earnest—and then answers himself: "At this point what do I believe in? *Nothing.* Who advises me? *Nobody* advises me. I'm all alone."

UNCERTAINTY

My family was scared when I went in for my thyroid
operation. They didn't know if it was cancer, or if the
doctors could get it all, or what. I think the feeling of
most people involved with atomic blasts is confusion.
Not being able to pinpoint the cause or what's going to
happen, not being able to find anyone who will take
responsibility. It's a vague uneasiness.

—**Laura Martin-Buhler,** Utah resident in the 1950s

Unresolved ambiguities compound the challenge technology survivors face. "What do I believe in? Nothing. Who advises me? Nobody." Psychiatrist Richard Lieberman points out that experiencing
the world as unpredictable and beyond one's ability to interpret is
a common human condition, yet it is not under normal circumstances a constant one.[1] Situations usually change. A semblance of
predictability returns, and one can refurbish the all-important assumption of security.[2]

The uncertainties posed by technology-induced illness, however,
present a special situation. For the most part, they are "static"
uncertainties: They can go on indefinitely with no resolution. There
can be a multitude of them, contesting one's sense of identity and
strength from every conceivable angle. Complicating the challenge
is the fact that this predicament is unfamiliar and unwanted. Having
been harmed in innocence or ignorance, the survivor is rarely prepared for a life of unrelenting uncertainty. Nonetheless, he or she is
faced with a host of unanswered—and often unanswerable—questions.

For many there is uncertainty about the reality of exposure. *Did
it actually occur?* Whereas the presence of a technology like an
intrauterine device is known to its user, the presence of many of

today's technologies is not. They are often deployed by agents other than oneself, like lumber corporations spraying forests with herbicides, and by agents unwilling to admit the existence of danger, like the United States government covering up leakage at nuclear weapons plants. To make matters worse, many technologies that threaten human health—from electromagnetic radio waves to pesticides—are invisible.

Andy Hawkinson reveals uncertainty about his exposure to radiation when he describes the South Pacific paradise of Eniwetok with its gentle breezes, blue lagoons—and no hint of danger. Ionizing radiation consists of subatomic particles that may hit and damage one cell of the body and not another, may lodge in an organ or pass straight through it, and when airborne, may settle into a community or pass right over it.[3] To Andy Hawkinson, not knowing if he was exposed to radiation is the most troubling aspect of his life. "Today I still don't know," he says. "*I can presume.* But there's not a person that can give me the answer. All I know is that I've suffered."

Likewise, Harriet Beinfield feels confused about her exposure to the herbicides sprayed around her home near California's Klamath National Forest. She *knows* she used the Dalkon Shield, and she *knows* this harmed her reproductive system by giving her immediate pelvic problems, but invisible gases wafting across the countryside present different issues. Harriet's son Bear was born with Holt-Oram syndrome, a developmental birth defect that produced holes in his heart and caused him to be born without thumbs. He is also missing muscles in his upper back. Harriet researched the relationship between local health problems and the hazards of the herbicides 2,4-D and 2,4,5-T for a community lawsuit to halt spraying in the national forest. She documented one case in which a family's water source was directly contaminated, and their child was born with medical defects identical to those of laboratory animals exposed to the same chemicals. The child of one of Harriet's friends was also born with the normally rare Holt-Oram syndrome. In one year twenty-three women in the community became pregnant. Eleven either suffered miscarriage or had children with birth defects.

Despite these facts and her own intuition about the situation, Harriet still feels doubt about her exposure. "If I have to explain Bear's condition, I'd say the possibilities are infinite," she says. "I'm sensitive to caffeine. Even Chinese tea makes me agitated. Then there are low levels of radiation that could have affected my genes. It's not as if I was around a nuclear test site or am part of an epidemiological group that's been exposed, but it's possible I have been. I took birth control pills. There are any number of channels for the mutation. Really, who knows if I was contaminated by herbicides?"

Then there is the question concerning the events leading up to exposure: *How did it happen?* The answer can be crucial for a survivor's legal plea for medical benefits from insurance companies or government agencies or financial compensation from the party deemed responsible. The answer is also crucial for peace of mind. It can be difficult to determine because the more complex a technological system is, the harder it becomes to determine which part of it went awry or made the impact. When Jane Woolf became ill with thyroid, liver, and immunologic disorders and she and the owners of the property she was caretaking in Michigan realized the water supply was contaminated, they immediately tried to figure out what had happened. By contacting state authorities, they learned that a gasoline leak had occurred ten years earlier at a gas station across the highway. They had their water tested, and indeed it contained the gasoline-based carcinogen benzene. But they were left with a sense of helplessness because the story was incomplete. They could not pin down how the leak had contaminated the water supply. "Perhaps the gas got in the well and started corroding the pipes," Jane guesses. "Maybe that's why the water tasted like metal. Then maybe the corrosion taste was so strong it covered up the taste of gas. There are all these theories, but nobody knows for sure, and we'd like to know. We'd really like to know."

Some technology survivors, particularly those affected by invisible contaminants like radiation or chemicals, are faced with uncertainty about the extent of exposure. The question is: *How much did I take in?* If a concrete answer is available, a survivor has more certainty in determining health effects and can approach parties deemed responsible with confidence. Documentation of a high dosage can provide some assurance that existing health problems are related to it and that those appearing in the future will be related to it. When exposure has been low, a person may stop worrying about diseases that could arise in the future. Too often, though, the exact level of contamination is difficult or impossible to determine—or the information is not within grasp.

When Jose Luis Roybal participated in the American "cleanup" of Hiroshima in 1945, monitoring technology was but a few months old. He asserts that measuring devices, when employed at all, were too primitive to be reliable. Jose Luis's claim is validated by a recent government controversy. For years the Pentagon contended that 99 percent of atomic veterans received no more than five rems of radiation, about the amount in three hundred X rays. The validity of these estimates was recently challenged by a General Accounting Office study examining the 1946 Operation Crossroads tests in the South Pacific. The study concludes that Pentagon estimates were not arrived at by strict scientific methodology. They do not account for the radiation that people inhaled or ingested and do not consider the high margin of error in early monitoring technologies[4]—all of which are factors making accurate measurement impossible.

In some cases, there is uncertainty about dosage because no one—whether out of naïveté, denial, or fear—has ever made the necessary measurements. The exact levels of toxicity in the Love Canal community were unknown when Hooker Chemical Corporation sold the landfill to the Board of Education in 1953. According to Lois Gibbs, Hooker had records of the twenty-two thousand tons of chemical waste buried and the board knew full well that chemicals resided beneath the school they built, but until residents became upset in 1978, officials never monitored leakage into the water, air, or soil.

For Vietnam veterans exposed to Agent Orange, the story is the same. The Air Force estimates that 17.4 million gallons of herbicides were used in South Vietnam and Laos between 1962 and 1971, including 368 pounds of one of the most dangerous forms of dioxin, 2,3,7,8-tetrachlorodibenzo-para-dioxin (TCDD), figures that Centers for Disease Control spokesperson Richard Diefenbach claims are low.[5] Throughout the war, the government made no attempt to measure how much of the herbicide actually contaminated soldiers. Today an increasing host of veterans in deteriorating health want to know if their diseases were caused by the defoliant—and want sanctioned medical benefits for their ills. Since the past cannot be reconstructed, the issue now becomes measurement of present levels of toxicity.

The United States government does not want to take such measurements and does not want anyone else to do it. According to psychologist Paul Scipione of the New Jersey State Agent Orange Commission, if the government can stave veterans off until the mid-1990s, the dioxin in their blood and fatty tissue will have been excreted. Dioxin-caused diseases will be unprovable, and admission of fault will be a moot point. The government appears to be executing this strategy. It refuses to allocate money for studies that would link exposure levels to illness, and instead it conducts research like the Air Force "Ranch Hand" study of pilots and crews who flew spraying missions, a study that may not be concluded until the year 2002. By that time dioxin levels in surviving veterans will be impossible to detect, and most veterans who currently have Hodgkin's disease, lymphoma, or the neurological disorders thought to be associated with dioxin will be dead.

In the meantime, states organizing their own studies have had to struggle for modest commissions against lobbying efforts by powerful chemical companies. Despite obstacles to its work, however, the New Jersey State Agent Orange Commission recently pioneered a scientific technique for measuring dioxin levels in the blood. Yet when they requested "an evening a month" or "a weekend every three months" to conduct tests in federal research facilities on equipment they could never afford, the government flat out refused.[6]

Difficulties in knowing exact levels of exposure may also result from political maneuvering at the time of exposure. An agency has indeed made measurements but is unwilling to release the information. In 1953, Frank Butrico was a Public Health Service Radiation Safety monitor measuring fallout from the Nevada Test Site in St. George, Utah. Frank reports that before a test called Dirty Harry, Atomic Energy Commission officials instructed him to give inconclusive reports of radiation levels to curious residents. "We are getting inquiries. Some people have gotten sick," one official told him. "Let's cool it, quiet it down. If we don't, there might be repercussions, and they might curtail the program, which, in the interest of national defense, we can't do." After the test, Frank says, his "instruments went off the scales." When he later returned to Nevada headquarters, he discovered that officials had *lowered* the radiation readings on his report.[7]

Similarly, in a 1983 court case atomic veterans brought against the then Veterans Administration for refused disability claims, veterans' lawyers subpoenaed VA records containing documentation of exposure levels. Despite falsified records and difficulties in measuring radiation, the documents must have been incriminating. The VA risked punitive actions by shredding the documents rather than producing them.[8]

91

Another disturbing question technology survivors face is what Henry Vyner calls "etiological uncertainty."[9] *Did exposure to a technology cause my illness?* According to downwinder Laura Martin-Buhler, "There's always the ambiguous feeling. Would I have gotten the illness anyway if I hadn't lived downwind of the test site? It's a constant sense of doubt and confusion." Nagasaki veteran Gilberto Quintana expresses a similar sentiment. "The money doesn't matter to me. The only thing I would like to know for sure is if my diseases were caused by radiation or not. It would make me satisfied inside to know."

Latency periods between exposure and illness contribute to this nagging doubt. Some cancers can take twenty or thirty years to manifest. Also many diseases are linked to not one, but to any number of technological exposures. They can also be linked to genetic, life-style, and psychological factors. One example is cancer. Cancer can be caused by exposure to the ionizing radiation of a plutonium spill in a nuclear laboratory. It is also suspected to result from the nonionizing radiation of radio waves,[10] dioxin in herbicides and Agent Orange,[11] and the electromagnetic fields generated by high-voltage power lines.[12] To add to the morass of influences, people who eat a great deal of fat can be more likely to contract cancer,[13] as can people who have recently endured the death of a loved one.[14] The question then becomes: Can a nuclear lab worker be absolutely certain his cancer is the result of plutonium exposure?

The problem is complicated by multiple exposures. Benina Berger Gould claims that she doesn't have "a technologically clear story." Benina used birth control pills, the Copper-7 IUD, and the Dalkon Shield. Resulting hormonal imbalances caused one doctor to think she had endometriosis, a buildup of uterine tissue outside the uterus that can cause cysts. He administered estrogen treatment. Benina began to undergo early menopause with constant abdominal pains and emotional imbalances. A dangerous cyst led to a complete hysterectomy. She was given estrogen replacement therapy at what another doctor later determined to be twice the necessary dosage.

In 1986, Benina discovered a lump in her breast. Could she say that birth control pills caused early menopause, endometriosis, or this lump? Could she say that the lump came from estrogen replacement therapy? Or from all of the prescribed technologies acting synergistically? No clinical or epidemiological studies exist to give Benina proof for a court case against the appropriate doctor or pharmaceutical company—or to give her the satisfaction of knowing.

Questions about cause are also exacerbated by a lack of clinical or epidemiological proof connecting a particular technology to a particular condition. Paul Scipione reports that Vietnam veterans exposed to Agent Orange endure agonizing perplexity about proof. Many health effects of TCDD have been minimized by government studies, and so veterans in ill health live without official acknowledgment—and without knowing for certain if their physical ills derive from exposure.[15] This same problem extends to most of the better-known incidents of contamination. Arguments still rage about whether the Three Mile Island accident caused an increase in neonatal hypothyroidism and infant death.[16] Any link between atmospheric nuclear blasts in Nevada and childhood leukemias in Utah is still not officially recognized.[17] In Michigan the conflict goes on over whether PBB contamination of livestock resulted in human illness;[18] and Love Canal residents still, despite their own efforts at epidemiological research, cannot point to an officially sanctioned study proving a link between chemical exposure and disease.[19] While they suspect either nearby communications towers emitting electromagnetic radiation or nuclear waste discharges from area hospitals, residents of San Francisco's Noe Valley have no sanctioned studies to explain an alarming increase in childhood cancers and birth defects.[20] Many technology survivors reside in a netherworld. For all their understanding of their own health histories, they cannot point to a supportive study showing the link between technology and illness.

Then there is uncertainty about the nature of one's illness. *What is wrong with my body?* I have discussed the limitations of Western allopathic medicine in diagnosing physical conditions, and Henry Vyner mentions the existence of "new diseases" caused by modern technologies and not yet named, studied, or understood. In recent years we have become aware of AIDS, chronic fatigue immune dysfunction syndrome, candidiasis, environmental illness, and electromagnetic sensitivities. Technology survivors often do not have any idea what illness they have, and, as one atomic veteran puts it, "to get a bad diagnosis is better than receiving no diagnosis at all."[21] With no sense of what is wrong, they do not know how to treat their condition or even which health professional to ask for help.

This problem is particularly pervasive among those exposed to invisible contaminants. After working at the GTE assembly plant in

93

Albuquerque for two years, Irene Baca lists the following disparate symptoms: insomnia; headaches; memory loss; fatigue; allergic reactions to smoke, perfume, and chemicals; facial paralysis; viral infections; mood swings; ulcers; a metallic taste in her mouth; arthritis; nervousness; depression; ringing in her ears; gum infections; and anxiety attacks. While GTE doctors denied that anything was wrong with Irene, a non-GTE doctor proclaimed she was entering early senility. Another non-GTE doctor, a neurologist, dismissed her experience, saying that in her thirties she "wasn't going to feel like [she] did when [she] was five years old." Irene's own family doctor finally broke down in tears, crying: "I want to help you! I want to help you! But *I just don't know how!*"

Uncertainty about the nature of disease leads to doubts about treatment. Yet another difficult question technology survivors face is: *What should I do?* After enduring what Carl Porter calls "the hundred doctors," the best direction to take becomes ever more ambiguous. Danny Morris was twenty-five years old when his dream of becoming a lawyer was squelched by the sudden onset of allergic reactions to pollens, animals, chemicals, wood resins, dust, and foods. Danny believes he was poisoned by pesticides routinely sprayed in his neighborhood in Cambridge, Massachusetts. He withdrew from Harvard Law School and began a confusing search for a treatment that could change his condition. Since 1984 he has tried drugs, allergy antigen treatments, special diets, the elimination of toxic materials from his home, and herbs—and yet his condition continues to deteriorate. He is weak, hardly able to function, and, in his own words, "sensitive to just about everything in the environment except porcelain."

Danny is about to begin a course of homeopathic treatment. "I don't know if it will help," he says. "Nothing has helped so far. I'm completely broke so I can't go on experimenting forever. This thing will be resolved one way or another. I'll get better or I'll die. So I *have* to try—even though I have no guarantee."

Perhaps the most challenging and universal question technology survivors face concerns the future. *What will happen to me?* Having been sickened already, many face the possibility of slow deterioration or a sudden onset of disease. As a DES daughter, Perry Styles has already experienced laser surgery for a precancerous condition of the cervix and two premature labors requiring her newborns to be confined in the neonatal intensive care unit. "Some days I wonder what will happen in the future," she reveals. "Will I get cancer? Will I have to go through more operations? Each time something like a negative Pap smear comes back, I wonder: Is this the first of many things? I need a couple of years without anything going wrong to start feeling better again. I don't even know if I'll get that."

Jane Woolf's concerns are similar. "I just don't know how I'm going to be. I'm optimistic a lot of the time. I think I can clear my body out using alternative methods the M.D.'s don't recognize. But I still don't know one hundred percent that I'm going to be okay. My immune system could deteriorate further. I could get leukemia. In the future I may not be able to work and what if I have kids by then? Last year I borrowed so much money I don't have anyone else to borrow from. The best thing would be if I just didn't get sick again. But with benzene poisoning you never know for sure."

Former naval shipyard worker Jesus Rives, exposed to asbestos on the job and diagnosed with asbestosis, makes the point succinctly. "You're a tough person. You joke. You cut up. But you live with the constant terror that tomorrow might be it."

The most obvious result of such unrelenting ambiguity is stress. According to Hans Selye, whether the stressor is physical or psychological, the body mounts a predictable response: a state of excitement whose purpose is to protect. If a person exists in this state indefinitely, though, trouble can result. Selye cites peptic ulcers, hypertension, and heart disease.[22] Other researchers point to psychiatric complications, depression, anxiety, and psychosomatic illness.[23] To Henry Vyner, "The uncertainty feeds upon itself, and it becomes the breeding ground for the remaining psychological effects of the contaminants."[24]

"Eventually," Vyner writes,

the exposed person learns that medical science will be unable to help him protect himself from the health effects of the contaminant. . . . He learns that he will just have to wait and see, that there is no way to empirically cope with and master the threats posed to his health by the exposure. At this juncture, the exposed person has two equally unattractive options. Either he can choose to forget—or deny—the possibility that his health has been compromised, or he can continue to attempt to pursue a vigilant course.[25]

Chronic ambiguities can also produce a state of traumatic neurosis. This condition usually occurs among people who have endured a major traumatic experience: a nuclear blast, an airplane disaster, or military combat. Recognized since the seventeenth century, the condition has been known variously as "shell shock," "disaster syndrome," and "traumatic stress." It was noticed widely among men who fought in World War I and more recently among Vietnam veterans. In 1980 the American Psychiatric Association sanctioned a new term to describe it: "post–traumatic stress disorder." Its symptoms include: (1) reexperiencing the catastrophic event in recollections, dreams, or feelings; (2) numbing one's responsiveness to the world; and (3) other symptoms which might include hyperalertness, guilt, sleep disturbance, or memory impairment.[26] The essence of the condition is that one continues to live as though a previous traumatic event were still occurring. It is hypervigilance become fixated in the personality.

We can certainly argue that many, if not most, technology events do not compare in catastrophic quality to a nuclear holocaust, an air crash, or war. They are more mundane experiences like walking down the street, eating an apple, or taking a pill. Yet these experiences do cause many survivors to go on living as if the initial incident were happening right now. "I realize how traumatized I am by this experience," reveals Susan Griffin. "I carry so much fear now. I go into states of anxiety over small things." "When Bear was born, he could have died," says Harriet Beinfield. "He needed surgery and twenty-four-hour-a-day attention, and we had to make life-or-death decisions. I was in a dull state of terror for five years." Sustained or delayed traumatic responses can be attributed to ongoing ambiguities. The offending event does not, in reality, disappear. Indeed, it continues to traumatize the survivor with real and potential health threats, real and potential financial stress, real and potential legal complications, and real and potential social failings.

Those who link their illness and a technological event know that because technologies are human-made, their suffering, doubt, and confusion did not have to happen. As a result of this awareness, they can become consumed with yet another unanswered question: *Why?*

"I tried to find a rationale," says June Casey. "The government knew how damaging the radiation would be. It's bad enough that they made the release, but now they should reveal the documents and tell us why. Even if they did something wrong, I would admire them, you know, if they told the truth. That they don't is something terrible. I have struggled to understand why. How could these men do this to us?! They say they were testing their instruments. That's the pat answer. But there's got to be more. I've looked to psychology. I've looked to history. I've looked into the human spirit. I've learned a lot, but honestly, I'll never know. *I will never know why they did it.*"

RESPONSIBILITY

We offered to settle for one dollar if the court would close the landfill. The court refused to close the landfill. They admitted that the landfill *did* damage the aquifer. It *did* poison my well. It *must have* poisoned my children. But we collected nothing. . . . We dreamed there was justice in the American courts. We got wiped out. You talk about David and Goliath! We were so naive to fight Shell Chemical, Ortho Pharmaceutical, Textron Incorporated, General Motors, General Electric, BASF . . . and they're a partial list of the people we filed suit against.

—**Frank Kaler,** New Jersey homeowner
whose well was contaminated with benzene,
carbon tetrachloride, lead, and radioactive waste[1]

One step beyond the question "Why?" lies another issue technology survivors must face. *Who is responsible for what has taken place?* Is it my fault for getting myself into the situation? Are multinational corporations, government, or the medical establishment to blame? Does accountability reside with the legal system? Or with society at large?

On one level, a technology survivor's attempt to come to grips with these questions is a psychological issue. One's personal beliefs about how the world should work and about moral behavior are at stake. Gilberto Quintana echoes the sentiment of many people when he asserts, "I just want justice!" Echoes Gay Ducey, "More than anything, we wanted the corporate people to admit they were responsible." On top of such moral challenges, though, there are concrete issues to consider. Who will pay for medical bills and lost wages? Who will compensate a survivor for suffering? Who will inform the

98

public about the danger? Who will prevent the event from happening again?

In their attempts to address these issues, survivors face the fact that technological society and the legal system regulating it function with a prejudice that prevents fair assignment of responsibility. When technological events harm people, two responses usually occur. First, responsibility falls to society at large. That is, you and I pay. The responsibility for technology-induced diseases—from asbestos and radiation exposure to environmental illness—is quietly diffused back to us by way of programs designed to divert public attention away from government and corporate roles in causing disease: raised insurance premiums, higher taxes, bankruptcy strategies, increased welfare rolls, and reduced government services. Unfortunately, we pay these expenses without knowing they spring from technology's excesses and without any democratic input into which technologies we are willing to risk our money for.

Second, the survivor pays. The responsibility for technology-induced illness resides with the individual made sick. Most survivors suffer alone, pay much of their own medical bills, and make do with a diminished capacity to earn a living. A common belief in democratic society is that "a person can obtain justice" and "the system works," but in fact there are legal, sociological, psychological, economic, and political factors that tip the scales of accountability away from the people and institutions deploying dangerous technologies— and toward the sick person.

The dynamic is reflected in many survivors' feelings about initiating legal action. They feel hesitant and afraid. They feel overwhelmed by the legalistic demands of conducting a case and doubtful that the arduous process will be worth the effort. Wendy Grace spoke with an attorney about her case against the makers of the Saf-T-Coil IUD. He felt she had a clear-cut case, but Wendy couldn't bear to pursue it. The task seemed herculean, and documenting the past decade of her life overwhelming. "The past is past," asserts Sukey Fox, rationalizing the fact that she will never be compensated for her suffering and loss. "I don't want to rehash old times. Let it lie."

Another common response is anger. "After my second operation, the idea hit me," says DES daughter Perry Styles. "I was sick of being the victim. The victim always had to pay. I would sue them!"

Both fear and anger are merited. A myth that has long captured the American imagination is that of the wronged Little Person who musters superhuman resources to win justice over the Giant, be it "the Company," "bureaucracy," multinational corporations, or the government. Films like *It's a Wonderful Life* and *Places in the Heart* and novels like *The Grapes of Wrath* tell the story that lies at the very foundation of the nation. While the ostensible theme of this story is the courage and triumph of the heroic individual who rises to the occasion, what the story assumes about American society is perhaps more revealing: It is dominated by extremely powerful and undemocratic institutions. In reality, it is rare that a person triumphs against these forces.

Waste management hydrologist Teddy Ostrow filed for workers' compensation. During the case Teddy's coworkers, in denial of the possibility that they too might become ill, claimed that chemicals do not harm health. He lost. Because he lives in California, a state that allows both workers' compensation cases and personal injury claims, Teddy could have followed the case with a second suit for personal damages, but feeling exhausted, alone, and discounted, he could not muster the energy.

The Ducey family, saddled with two chronically ill adults and a disabled son, settled out of court for $12,000 in their case against a major chemical corporation, barely enough to cover the expenses of one of their son's eight surgeries. As Gay wryly explains, "We knew we weren't entering the lottery."

Only 5 percent of product liability suits against technology manufacturers make it to court.[2] Of 3,050 claims filed by atomic veterans by 1986 for payment of expenses relating to their inability to work, the government approved less than 1 percent.[3] Of 7,781 claims for medical expenses in radiation cases filed by 1987, 4,979 were denied. Only 17 veterans received 100-percent disability, and only 1 of these men lived long enough to receive his benefits.[4]

Given these stories, the question arises: How is it that the Giant succeeds at walking away from responsibility so often?

On the surface of things, the legal setup can be perceived as fair. Tort law is supposed to right civil wrongs by determining responsibility between alleged victims and alleged perpetrators of harm. It appears to lay a foundation for mutual responsibility that, in a court of law, may be tipped one way or the other depending on the actions and knowledge of each party.

Several aspects of tort law reveal its attempt to approach issues of responsibility from a balanced perspective. For one, as part of a general liberalization in the 1960s and 1970s, class action suits are now acceptable. Residents of a neighborhood near a poisonous landfill, users of a harmful drug, or workers at a toxic factory can now sue the manufacturer of an offending technology.

Second, since 1965 the standard of strict liability has required manufacturers to be answerable for the effects of their products, stating that anyone "who sells any product in a defective condition unreasonably dangerous to the user or consumer . . . is subject to liability for physical harm thereby caused to the ultimate user or consumer." "Unreasonably dangerous" means "to an extent beyond that which would be contemplated by the ordinary consumer who purchases it."[5]

Strict liability also extends to occupational settings. It didn't always. In the early years of industrialism, occupationally caused illnesses were rarely recognized. English common law saw workers not as human beings with rights, but as part of the employer's machinery of production; that is, as his property.[6] Seventy to 94 percent of injury and death claims went uncompensated,[7] but since 1973, manufacturers are required by law to disclose workplace risks of their products.

The doctrine of proximate causation is, however, prejudicial against technology survivors, except in a few select cases. Originally developed in the nineteenth century to protect industrial owners, the doctrine made it necessary for a worker seeking compensation to prove that an employer's actions or technologies caused harm. In this century the ambiguity inherent to technology cases has been used for the benefit of owners and purveyors. How, for instance, can a worker *prove* that a molecule of PCB caused her cancer? How can a worker say *for sure* that he would not have had a heart attack anyway?

Owing to the recent wave of cases involving technology-induced diseases, though, the doctrine has been altered to make it possible for some survivors to gain compensation. In cases of asbestos-related illnesses and DES injuries, survivors are no longer required to prove who manufactured the product that caused injury, but only that (1) they have sustained injury; (2) the injury was caused by a toxic substance; (3) the defendant produced such substances; and (4) they were exposed to it. The burden of proof shifts to the defendant, who must disprove the survivor's claim or show that the particular technology they produced or disseminated did not cause the illness. Although asbestos and DES survivors still have obstacles to overcome, the burden of proof that falls to them is no longer impossible to meet.

To counterbalance liability of the manufacturer, United States law also holds that an individual is responsible when he uses a product knowing of its risks. This is the rule of informed consent. If the user knows about a technology's dangers and does not guard against them, he is guilty of negligence and cannot get compensation for resulting illness. If the manufacturer does not reveal the dangers inherent in a product, it is negligent and must pay the user for harm done.

The setup seems fair on the surface. In fact, though, other less visible factors are at work, factors that perpetuate a historical preference for the makers, purveyors, and financial benefactors of technological development. Such a prejudicial approach is hardly surprising in a society whose *raison d'être* is technological development.

The prejudice begins with the question of knowledge. How does a person know that a technology is harmful to health? A typical answer to this question is documented in one of the earliest cases brought by workers against asbestos-insulation manufacturers. The year was 1969, and the defense attorney was trying to get insulation worker Clarence Borel to admit he knew he was working with asbestos.

" 'I knew I was working with insulation,' Borel explained.

" 'Did you know it contained asbestos?'

" 'Yes sir,' said Borel. *'But I didn't know what asbestos was'* " (emphasis added).[8]

Despite legal standards defining the parameters of liability, an employer has no *immediate* economic incentive to reveal the extent of risk involved with machines or substances provided to do a job. "If you inhale these fumes, your kidneys could fail and you could spend the rest of your life on renal dialysis." "Just touching this substance could cause cancer." Despite the 1965 standard of strict liability and recently passed right-to-know legislation in the United States, a worker rarely hears such specific, straightforward warnings. To complicate matters, in some cases the employer isn't even aware of the dangers. The manufacturer is, and like the employer, the manufacturer has no economic incentive to reveal them and often does not.

Consumers encounter a similar vacuum of information. Thérèse Khalsa drank a protein drink containing NutraSweet, and indeed the label announced, as federal law designs it must, the ingredients of the sweetener: aspartame. But what is aspartame? To Thérèse it was a big word. She didn't know that ingesting it could lead to brain damage and neurological deterioration. The result was that Thérèse knew she was using something with the healthy-sounding name of NutraSweet. She even knew NutraSweet was made of aspartame, but she had no idea what it was.

Military personnel, mental patients, and prisoners used in human experimentation face a similar predicament. When they were ordered to witness nuclear tests in Nevada and the South Pacific, United States soldiers were not given enough information to even contemplate a choice between being irradiated or risking punitive action. Officers told them that indeed radiation was poisonous, but that they would never be close enough to receive harmful doses. "Watched from a safe distance," an army film told soldiers,

> this explosion is one of the most beautiful sights ever witnessed by man. The radiation may be high, but you will be moved out to avoid sickness. Finally, if you do receive enough gamma rays to cause sterility or severe illness, you'll be killed by glass, flying debris, or heat anyway. Don't worry about yourselves. As far as the test is concerned, you'll be O.K.[9]

Similarly, twenty-four-year-old Oregon State Prison inmate John Atkinson was enticed into consenting to participate in a medical experiment. He was told his participation would further scientific knowledge and he would receive fifteen dollars a month, a sizable sum for a prison inmate in 1963. He agreed. His testicles were subsequently and regularly bombarded with two hundred rads of X radiation, an amount equivalent to twenty thousand times that in a normal X ray. John claims that doctors in charge never told him that the experiments could produce burns, infections, tumors, sexual impotence, and birth defects—or that they would not let him out of the program unless he submitted to a vasectomy.[10]

In the case of residential communities exposed to toxic substances released from factories, laboratories, or waste dumps, residents have no idea that they are even being exposed, much less that they are at risk. People living near the Fernald uranium-processing plant in Ohio have been exposed to some 370,000 pounds of uranium dust released into the air and water and unknown amounts of seepage into the water table from storage pits and aboveground drums.[11] They were never warned. The neighbors of the Ponderosa Products particle board factory in Albuquerque didn't know the sawdust broadcast on their yards, cars, and sidewalks was coated with the carcinogen formaldehyde.

Sociological and psychological factors also contribute to a lack of forewarning. The burden of asking relevant questions falls to the individual who may have no inkling that questions are even in order. Reinforcing this setup are value priorities that further overshadow inquiry.

For one, the intimidation of authority can be a factor preventing people from gaining knowledge. In the occupational hierarchy, people on the lower rungs of authority are often afraid to ask questions. If they do, they could lose their jobs. Irene Baca felt such gratitude to have "a good job opportunity" at the GTE Lenkurt plant that she would never have questioned her superiors. Plus, having grown up on an Indian reservation, she knew nothing about chemical hazards. According to American studies scholar Stephen Fox, the innocence and lack of education of workers like Irene is a sociological factor GTE consciously builds into its production strategy. It is why GTE originally sited its plant in Albuquerque and why it later relocated to Juarez, Mexico.[12]

Likewise, many people feel intimidated by the all-knowing attitude that pervades the medical profession. "Do as the doctor says." In the face of such authority, enhanced by the power suggested by the presence of high technology, few people ask how a drug or device might affect their well-being.

Also disorienting to survivors is the fact that many sustain illness at the hands of the very institutions with which they identify. Former anesthesiologist Carl Porter feels incapable of expressing anger at the medical establishment and has never considered a legal suit for his sizeable medical expenses and lost wages. Yet a suit could be relatively easy to win. In the early 1980s, the medical world publicly admitted its responsibility for unnecessarily exposing operating room personnel to toxic chemicals, and a multitude of studies—including the all-important National Institute for Occupational Safety and Health/American Society of Anesthesiologists study—documents the health hazards of anesthesia gases.

Yet Carl is conscious only of what medicine has given him. "It took what had been a very insecure kid with low self-esteem," he reveals, "and gave me a way to get out of myself. It gave me all those personal rewards of having patients like me." He is only partly conscious of what medicine has taken away from him. Clinging to denial, even to the extent of burdening his parents with medical bills for over $300,000, he feels disconnected from other people made sick by operating room technologies, isolated, and confused about responsibility. He openly wonders, but cannot bear to answer: "What out there is picking on me?"

Then there is peer group pressure. When the Dalkon Shield hit the United States market in 1971, its massive publicity campaign in books, magazines, and medical journals merged with the baby boom generation's preoccupation with the "sexual revolution." The word circulating among women was that this birth control device was safe, trouble-free, and "the thing to do."

Lois Gibbs's husband faced peer pressure that militated against his curiosity about some odd facts: his skin turned from white to brown, and he perspired brown liquid into the bedsheets at night. At Goodyear Tire and Rubber in Niagara Falls, where workers handle and breathe a host of toxic chemicals, the overriding ethos among male workers was to "do the job like a man." According to Lois, "The guys at work dealt with the brown sheets and the chemicals oozing out of their pores in a making-fun sort of way, like 'The wife was really pissed off because her whites weren't the whitest.' "

What results from such influences are millions of people made sick by technologies without awareness of risk or conscious consent. A related burden that falls to them is that of documentation. Since few people are aware of risk at the time of use or exposure, they do not take the precaution of recording the exact circumstances or technologies involved. Later, in a court of law, this deficiency of information works against them. In his attempt to bring the first case against asbestos-insulation manufacturers for injury to workers, Texas attorney Ward Stephenson was required to prove the source of his client Claude Tomplait's illness. He spent long hours questioning Tomplait about the contractors he had worked for and what materials he had used. Stephenson also tried to garner information by interviewing Tomplait's coworkers and union officials, but in the end he lost. Tomplait could not remember exactly when and on which jobs he had used products made by the defendant.[13] When Anika Jans tried to put together a case against an oral contraceptive manufacturer, she could not even prove she had used the Pill. In the

105

ensuing years Planned Parenthood had thrown away her files, and she had not thought to keep records.

Interweaving with the problem of documentation is that of the statute of limitations. Ordinarily a person must begin litigation within two or three years of the date of the cause of injury. Many technology-induced illnesses, however, develop imperceptibly for years before they manifest. Since many survivors do not understand the connection between a technological event and illness, new questions arise that push the limits of ordinary tort process. Does the statute of limitations clock start when the technological event takes place? Does it begin when the person suspects a technological event caused injury? Or when an official study proves the connection?

There has been some attempt to address these ambiguities. By 1986, forty-seven United States federal districts had adopted statute of limitations laws that start the clock at the date of discovery of the cause of illness. In a Supreme Court case in Delaware in 1984, it was decided that the statute of limitations for injuries arising from asbestos inhalation began when asbestosis first became evident, not at the time of exposure or when scar tissue began to form.[14] Such developments are helpful in rebalancing the issue of responsibility for technology-induced illnesses, but laws remain fragmented across the country and from technology to technology.

When survivors do know enough about the connection between a technological event and illness to take their case to court, they find themselves at an economic disadvantage. Most purveyors of dangerous technologies are corporations and government agencies, and most technology survivors are individuals. Attorneys are expensive. Sukey Fox knew she had a case against A. H. Robins, but she was earning $625 a month and could not pay for legal services. Typical of low-income people, Sukey did not know enough about the legal process to know alternative payment approaches existed.

In one such alternative, survivors band together to attract socially conscious lawyers willing to give of their time. Until 1988, veterans of the armed services were barred from hiring attorneys by a Civil War-era law stipulating they could not spend more than $10 on a lawyer. As a result, atomic and Agent Orange veterans relied solely on donated services in their claims against the Department of Veterans Affairs.

Other survivors find lawyers willing to take the chance that cases will win and offer their services on a contingency basis. These are, as asbestos worker Jesus Rives says, "lawyers working for the people." Ward Stephenson was one of them. Having learned of the injustices asbestos workers face, he spoke at a trial lawyers' seminar in 1969 declaring that survivors were "cry[ing] out for help from members of the bar" and that the legal professional could render "a substantial benefit to mankind" by alerting the public to the dangers of asbestos.[15] As technology-related litigation increases, law firms specializing in particular technological events emerge. Some are large firms that invest in the research necessary to handle DES, asbestos, or toxic disposal cases. Then there are lawyers at small firms who feel, as did asbestos-litigation pioneer Stephenson and GTE workers' counsel Josephine Rohr, an outrage that propels them to dedicate time and energy against great odds.

And the odds *are* great. According to Bruce Berner, deputy prosecutor in the Ford Pinto case, "In corporate criminal cases the prosecutor is often twice cursed—once for having fewer resources and once for playing by rules which assume it has more."[16] In order to present a solid case for a technology survivor, an attorney must spend an enormous amount of money to present studies, medical tests, legal research, and expert witnesses equal in caliber and credibility to those that a large corporation can produce with ease. A medical doctor or psychologist may charge $150 an hour for consultation, a medical witness $1,500 a day for testimony. A study can cost thousands of dollars. If the attorney is part of a law firm, that firm must front the money for these expenses. If the attorney is state-hired, she must justify the expenditure to the constituency and convince related regulatory agencies to cooperate.

On the other hand, corporations have whole departments of attorneys financed year-round to cover the bases for possible litigation, plus academic and medical expert witnesses often paid not just to testify, but also to do studies on behalf of the technology in question. The eight volunteer lawyers working for a National Association of Radiation survivors case to abolish the $10 fee limitations were up against twenty-seven lawyers hired by the Justice Department. A. H. Robins retained a host of Ph.D.'s and ob-gyns as witnesses and consultants, paying them, either directly or through other agencies, in some cases hundreds of thousands of dollars.

Other roadblocks also stand in the way of technology survivors' obtaining justice. Established in 1950 by the Supreme Court as a protective measure for the government, the Feres Doctrine prohibits employees and citizens, in most cases, from suing the government for its activities. As a result, Utah residents downwind of nuclear testing have been unable to bring suit for resulting cancers. Likewise, atomic and Agent Orange veterans cannot sue the government and must instead bring claims for medical care before the Department of Veterans Affairs.

In 1984 Congress adopted the Warner Amendment, prohibiting United States citizens—whether atomic veterans or downwind residents—from suing federal contractors who manufacture bombs or radiation testing gear. A class action suit against AT&T Bell Laboratories, the University of California, and other government contractors by the families of servicemen exposed to radiation in Nevada, the South Pacific, Hiroshima, and Nagasaki was then denied in federal court in 1986.

Such a climate of political resistance is echoed in the occupational setting. According to state workers' compensation acts, workers cannot sue their employers. Enacted between 1911 and 1940, these consisted of industry-supported legislation to diffuse pressure from workers injured and killed in coal mines, textile factories, and steel mills. On paper, workers' compensation appears fair. It takes away the employees' right to sue for full compensatory damages in exchange for making employers liable for work-related injuries and death regardless of whether there has been negligence. From the workers' point of view, though, it is not fair. Workers' compensation allows employers to compensate a worker for loss of health or life at amounts *much lower* than would be required in court and at amounts that neglect to take into account such factors as a worker's medical expenses and loss of income *over a lifetime.* The system also eliminates large punitive awards that might serve as incentives for industry to remedy hazardous work conditions.

An example of the effects of workers' compensation can be seen in the workers' case against GTE Lenkurt in Albuquerque. Although a court order prohibits disclosure of the amounts awarded, it is widely believed that workers like Irene Baca and Susan Hernandez, whose health was ruined for life, received between $15,000 and $30,000.[17]

Irene's feelings about this situation are strong. "I am bitter," she says. "I am bitter because I don't think anybody has the right to expose you like that and take your life from you and rob you of your livelihood. If we have to spend the rest of our lives—however many years that is—I feel we need to let the world, the United States, and the state of New Mexico know what these chemical companies are getting away with. They made millions and millions of dollars off our work, and now all we ask is that they compensate us for what they've done to us, for how they've taken the ones we love away from us. And they don't want to do that. It's not right."

In their attempts to avoid paying compensation for technology-induced diseases, large corporations also invent complex measures to sidestep responsibility. In the face of increasing liability suits, the asbestos manufacturer Manville Corporation (formerly Johns-Manville) and the Dalkon Shield-maker A. H. Robins pioneered the Chapter 11 strategy: the declaration of bankruptcy with attendant freezing of all assets and refuge from lawsuits. By convincing the court that the potential liability of the company would require it to establish a reserve fund that would wipe out its net worth, these corporations entered into prolonged periods of reorganization, thereby calling a halt to thousands of lawsuits brought against them. Interestingly, in 1982 when Manville "went bankrupt," it ranked 181st on *Fortune*'s list of the nation's 500 leading industrial corporations—with assets of more than $2 billion.

A. H. Robins was not financially shaky when it requested protection in 1985. Rather, these actions are viewed by bankruptcy experts and survivors alike as strategies to avoid responsibility.

According to management specialist Arthur Sharplin:

> In terms of the personal interests of the managers who led Manville into Chapter 11, reorganization has proven an overwhelming success. For them things have turned out much better than they could have imagined. . . . The burdens of management have been lightened by the bankruptcy law giving Manville the opportunity to continue to bring in money and relieving them from having to pay their debts. One by one the five senior managers are leaving, with huge severance checks and pensions—not to mention the spiralling salaries they have gotten in the interim. Chairman McKinney, for example, retired August 1, 1986, with $1.3 million in severance pay and a pension of over $300,000 a year. . . . Moreover, the alleged connection of McKinney and others with tens of thousands of asbestos deaths and injuries has been obscured by time and astute public relations. . . . Disappointment is

hardly a strong enough word for the despair and outrage asbestos victims and their families feel. An estimated 2,000 of those who were seeking compensation in August 1982 have since died. In a remarkable irony the asbestos victims will find themselves sharing the Trust money with Manville codefendant asbestos companies, who have claims against Manville because they had to "go it alone" in defending asbestos lawsuits after the Chapter 11 filing. Money from the Trust will also be used to defend Manville, past and present Manville executives and directors, and Manville's insurance companies. The Manville managers have done O.K. Dozens of defense attorneys have made millions. The banks expect to be paid in full, with interest. And the asbestos victims? Those who remain alive wait for what will surely be less—and later than they can possibly imagine."[18]

Or as Heather Maurer says, "Manville has to pay attention to people who get up and scream and yell. But when they're in the ground and dead, they don't."

Review agencies like the Occupational Safety and Health Administration (OSHA) and the Environmental Protection Agency (EPA) were established, in part, to ensure effectiveness of the regulatory process and to ease litigation loads on courts. For the technology survivor, though, they often function as yet another legal roadblock to justice. In many cases, the survivor must exhaust costly administrative remedies before he can sue, and in most cases, he faces regulators who, as former corporate executives, favor unlimited technological development.

On the federal level, OSHA exists to set standards for work safety. Its role is to inspect workplaces, make citations for violations, and propose penalties, and it can seek federal injunctions to shut down offending operations. Ever since its inception in 1970, though, OSHA has been ineffectual. Pressured by corporations and compliant administrations, it has suffered from low enforcement budgets, appeals favoring employers, and low fines. During the Reagan administration, fiscal cuts brought additional restraints against thorough inspections, enforcement of safety standards, and issuing new standards, and made OSHA, according to former labor secretary William Brock, "impossible to run a good program."[19]

Also during the Reagan years, the EPA discontinued a key program for measuring the accumulation of toxic substances in the human population. The National Human Adipose Survey had provided the only long-term measure of absorption of chemicals. It revealed that virtually all Americans have PCBs stored in their body fat, and this discovery played a major role in the banning of the production of PCBs. The program also showed that human levels of the pesticide DDT declined after it was banned.[20]

Meanwhile, the EPA Superfund program, established to clean up toxic waste sites, limps along with inadequate enforcement, questionable treatment methods, and notoriously poor management. During the tenure of EPA administrator Anne Buford and Superfund director Rita Lavelle, only six cleanups were completed—and most of those were treated simply by capping contaminated wastes in place or hauling them off to landfills where further contamination could occur. In 1986 a congressional act set forth a 1991 deadline for cleanup of 375 of the 1,077 identified sites. By 1989 the Superfund had begun only 55 of these.[21] And the Superfund, established to help people harmed by toxic events who have no legal redress, contains no provisions for compensating survivors.

On the state level, technology survivors face a political struggle at every turn. Right-to-know legislation is supposed to guarantee workers and communities access to information about toxics they are exposed to, yet these laws have been systematically watered down as a result of effort by industrial lobbyists like the Semiconductor Industry Association in California and the Massachusetts High Technology Council. The New Jersey State Agent Orange Commission has had to lobby fiercely against chemical companies to receive its meager budget.

Then studies are often politically motivated and determined. Much of the research on the health effects of radiation is controlled by government agencies with a vested interest in nuclear power and weapons. Although some government data shows that the dangers of radiation are greater than current protective standards assume, the information is misrepresented or withheld, and independent scientists, such as those who documented cancer rates among workers at the Hanford Nuclear Reservation, are subjected to smear campaigns.[22]

Falsification of study results is common. A case in point is that of the Hazard Evaluation System (HES) of the California Department of Public Health. When the Mediterranean fruit fly descended

111

on California citrus crops in the early 1980s, farmers and chemical companies wanted the state to eradicate them with the pesticide malathion. Since it was a suspected carcinogen, HES director Marc Lappé did a study. He concluded that malathion may be safe for adults but that *its effects on infants, children, and pregnant women are uncertain.*

A series of political maneuvers ensued. The government made a statement using paragraphs from the study out of context to support malathion spraying. The health department gained a gag order on Lappé to keep the study from the press. National Toxicology Program researcher Melvin Ruber was fired after he demonstrated the link between malathion and cancer, and a birth defects registry was established in the heart of the sprayed territory, but it did not employ existing methods to detect defects caused by malathion.

Then, in a strategic move, the department promoted Lappé to a permanent toxicology position in state government—provided he never again try to influence policy decisions. Lappé resigned. His malathion study was labeled "Draft," and a new one was done. Its conclusion: Malathion may be safe for adults, and *there is no evidence to show toxic effects on infants, children, and pregnant women.* [23]

As if all of these blocks to justice were not outrageous enough, the final blow is personal harassment of technology survivors by institutions on the defense. The techniques of A. H. Robins, manufacturer of the Dalkon Shield, have long been known. During a deposition in Minnesota, the Robins counsel directed a Boston woman to answer questions about which way she wiped herself and how often she engaged in oral and anal intercourse. According to attorney Kenneth Green, questions they asked his client about her sex life ten years *before* she was even fitted for a Dalkon Shield were "disgusting as well as irrelevant." [24] Diane Carter describes her time in deposition as "being treated with the same disdain they give a rape victim."

Gay Ducey's experience with one of the nation's wealthiest chemical companies is comparable. According to Gay, the company's lawyers subpoenaed them on Christmas Eve. Once they ignored a restraining order prohibiting access to Gay's psychotherapy records, sending a messenger to the clinic to convince a secretary to hand them over. Seven-hour depositions were taken in a company room not only bugged, but equipped with a company-hired recorder trained to read upside down across the table and copy every word

Gay, her husband, and their lawyer wrote in notes to themselves and each other.

The most painful moment of harassment took place in one of these depositions. "It was a long and arduous process," Gay recalls, telling a story that encapsulates the feelings of many technology survivors who attempt to address the question, Who is responsible for what has taken place? "The lawyer had been baiting us as much as he could. It was a real war of wits. But then the guy said, 'What's your son's scar like? Is it a little scar? It's a little scar, isn't it?' . . . Well, I was so consumed with rage that I had to stop. . . . You see, there is no small scar on this kid of mine. *It's a big scar, man. It's a big deal!* To turn that kid over to surgery after surgery, to know that dioxin has harmed him for life, to know he's always going to have a hard time. I was *consumed* with rage! I almost struck the guy! It was a very august affair, lots of three-piece suits and I was in my best bib and tucker. I almost reached over, and I felt both my husband and my attorney, each one of them grabbing an elbow to restrain me from going right across the table and punching the guy out."

Who is responsible? Faced with blocks against an honest approach to this question from every conceivable front, technology survivors come to new questions: Why do parties who do in fact bear responsibility for technology-induced disease resist admission so intensely? And how do they get away with it?

THE
TABOOS

Taboos are very ancient prohibitions . . .
concern[ing] actions for which there existed
a strong desire.

—**Sigmund Freud,** *Totem and Taboo*

WE DON'T KNOW
THE WOUNDS

There's a kind of blindness that's based on fears, that's based on ease, that's based on lack of sensitivity and perception.

—**Rhiane Levy,** sensitized from overexposure to electromagnetic radiation in computers, fluorescent lights, TV and radio waves, and other electronics

The many wounds endured by people like Gay Ducey, Diane Carter, and Carl Porter are painful to contemplate and devastating to live. They are also repeated all over the world every day. Yet as individuals and a society, we manage to avoid knowing about them. Important questions arise: How many stories like these are aching to be told? How many nuclear industry workers have succumbed to cancer and leukemia? How many workers in microchip production are suffering from immunologic deterioration and are struggling to get financial compensation? How many DES children are afraid to have children of their own? How many people were contaminated, and continue to be contaminated, by the Church Rock, New Mexico, radiation spill of 1979?

When I set out to research the psychological consequences of technology-induced illness, I decided to make a list of how many people in the United States have become sick by exposure to or use of some of the most dangerous technologies. I started by calling the National Institute for Occupational Safety and Health. After explaining my goal to four department heads in three states, I was sent empty-handed to their Bureau of Statistics. They sent me to a separate agency, the Technology Data Center. No information there. Then I tried the Consumer Product Safety Commission, the National

Technology Information Service, the Centers for Disease Control, the Bureau of Labor Statistics, the Office of Technology Assessment (OTA), and the Environmental Protection Agency (EPA), several of which sent me back to the Technology Data Center. Still with no numbers, I turned to a host of private organizations like Asbestos Victims of America and the Environmental Law Institute, and then to the public health departments of some of the nation's most distinguished universities. I looked long and hard.

I learned that there are legitimate reasons why exact numbers are hard to find. For one thing, many technologies that can threaten health have not been officially recognized as dangerous. In the case of chemicals, a prime source of this deficiency resides in the sheer numbers of chemicals invented and disseminated. Some fifty thousand synthetic substances exist, and about one thousand are added each year. Minimal government and industry regulation results in places like Silicon Valley, an area south of San Francisco where high-tech manufacturing has poisoned land, air and water, or Cancer Alley in Louisiana where the air and groundwater are contaminated with industrial solvents because standards never hindered their use. According to the National Academy of Sciences, "Of tens of thousands of commercially important chemicals, only a few have been subjected to extensive testing, and most have scarcely been tested at all . . . [yet] it is clear that thousands or even tens of thousands of chemicals are legitimate candidates for toxicity testing related to a variety of health effects."[1]

In such an environment, knowledge of the effects of synthetics on human health is woefully inadequate. The EPA conducts much of its research by dosage monitoring, outfitting testers in vests laden with instruments to measure how much contamination a person encounters in a day; while university, industry, and government researchers use the laboratory model many of them admit is fallible, injecting animals with chemicals to determine their health effects.

The problem is aggravated by varying interpretations of the available data. Politically motivated arguments ensue between those with a vested interest in allowing a technology to be used and those who fear its effects. As an example, the federal government did not release its annual list of cancer-causing substances in 1988 because industry lobbyists quarreled with the methods used to attain the list.[2] Likewise, controversy over the health effects of video display terminals rages on. Industry and government research tends to focus on indi-

118

vidually correctible problems like eye strain and backaches, while independent researchers point to systemic problems like the health effects of the electromagnetic radiation, ELF waves, and microwaves the machines emit.[3]

Exact numbers of medically affected people can also be hard to determine because the sources of their exposure have not been located. For instance, the OTA estimates that there are some six hundred thousand toxic dump sites in the United States. The EPA has designated over two thousand of them for priority cleanup, while Congress's General Accounting Office insists that the total needing immediate cleanup could be double that figure, and OTA reports that the number could exceed ten thousand sites.[4] Then there are the uncounted, and perhaps uncountable, sites where wastes have been illegally or inadvertently dumped.

In addition to these barriers to counting exact numbers of people made sick from exposure to technologies, there are others. There can be difficulties in getting medical diagnosis—and so difficulties in documenting affected populations. Many survivors do not get adequate diagnosis of their illnesses, and latency periods of many diseases complicate this problem. Also, there is a vast range of individual medical responses to any given contaminant. Finally, some toxic technologies may not cause obvious disease during the lifetime of the person exposed, but rather alter that person's genes and affect later generations.

I had begun this research wanting to reveal what I considered important news: the vast numbers of human beings made sick by dangerous technologies. At this point, though, I could see that the news could no longer be those numbers. It was something worse. The news was the fact that the numbers do not exist.

Aside from all the reasons it is difficult to make such a count, at heart, we do not know how many people are suffering because our society has not made an integrated effort to assess the human consequences of its technologies. The onus of this predicament lies with the scientists and engineers who sidestep responsibility for the social uses of their inventions. It lies with the government officials and university researchers who shield their minds from the health impact of the technologies they espouse. It lies with the military professionals whose drive to propagate war technologies cancels any concern for the human implications of their creations. And it lies with the corporate executives squelching and shaping public information to enhance the economic benefits they derive from the technologies they launch into society. The problem originates with these sectors of the population, but its perpetuation is fueled by more complex forces. We cannot ignore that, on the whole, citizens allow themselves to be no more concerned or informed than the institutions that either disseminate dangerous technologies or prevent us from knowing their inherent dangers.

The problem is not just that many of us—from citizens to scientists—do not acknowledge the dangers. It is that we do not allow ourselves to admit that our neighbors, family members, and even we ourselves may be suffering from technology-induced illness. We have technology taboos to protect us from this perception, agreed-upon rules and unconscious restrictions we learn through socialization and that speak to our deep need to avoid certain experiences. There is a taboo against challenging our technology, there is a taboo against questioning the institutions that purvey our technology, and there is a taboo against confessing harm by technology.

The sociologist Jacques Ellul suggests how such a system of taboos functions. In his *Propaganda: The Formation of Men's Attitudes,* Ellul sees the determinant of public perception as more than indoctrination thrust upon the population by a cabal of self-serving officials and executives. He sees it as a system, a partnership, with all sectors of the population involved. "There is," Ellul proposes, "a citizen who craves propaganda from the bottom of his being and a propagandist who responds to this craving." The root of this relationship can be traced to the nature of technological society. For the first time in the history of humankind, people are uprooted from the traditional support systems of family, community, nature, and spirituality. "That loneliness, inside the crowd," writes Ellul, "is perhaps the most terrible ordeal of modern man . . . [and] for it . . . propaganda is an incomparable remedy."[5] Simplified, concrete images purveying technology's wonders supersede the complexities and uncertainties created by living in a fast-changing technological world.

What we have in modern society is a set of technology taboos that directly benefits, at least in the short run, the creators and disseminators of technologies. What we have are taboos that indirectly satisfy the psychic needs of the general population with their promise of "the good life," glamour, and "progress." What we have is a whole society unaware of the terrible fact that Gay Ducey, Diane Carter, and Carl Porter have been wounded.

120

PSYCHOLOGICAL TABOOS

The range of what we think and do
is limited by what we fail to notice.
And because we fail to notice
that we fail to notice
there is little we can do
to change
until we notice
how failing to notice
shapes our thoughts and deeds

—**R. D. Laing,** "knots"

Technology taboos are set in motion, in part, from a psychological matrix: Defense mechanisms like repression, denial, projection, introjection, personal disconnection, rationalization, and selective inattention. Such mechanisms act as filters of attention that allow humans to avoid uncomfortable experiences, and they are built right into the structure of our minds. Anthropologists and psychologists suggest that these mechanisms originated for some very good reasons: to enhance social cooperation, to enable us to avoid unmanageable pain, and simply to help us sift through the barrage of stimulus bombarding our senses. Ultimately, they probably came about to keep our attention free of the terrifying fact that we are frail and vulnerable beings and that—somehow, someday—each of us will die.[1]

Originating in the human mind in order to facilitate sanity and survival, when they are transferred to the issue of modern technology, defense mechanisms become destructive to both sanity and survival. They keep us from awareness and block our access to psychic resources that could serve sanity and survival. We might liken them to a snake biting its own tail: As we defend our minds from questioning our society's technologies, so we prevent ourselves from knowing what these same instruments are doing to our health. As we prevent ourselves from knowing what technologies are doing to our health, so we defend our minds from questioning them.

Unconsciously circling through this dynamic, we find ourselves enhancing the cohesion of a society whose *raison d'être* has become technological development, and so we collude with the wounding of its citizens. We avoid the discomfort of being without modern technologies like cars and hair dryers or of transgressing accepted mores that equate technological development with human evolution. We ward off the magnitude of technological stimulus that comes our way each day—the blinking billboards, freeways, electronic music, and microwaves—and we keep our minds distracted from the fact that modern technologies are, in fact, causing us to become increasingly frail and vulnerable—and to die.

REPRESSION: Repression is often seen as the original defense mechanism and the one upon which all others are built. When we repress, we blank out the facts and we blank out our feelings about the facts. Psychologist Daniel Goleman sums up repression when he writes, paraphrasing R. D. Laing, "One forgets, then forgets one has forgotten."[2] We repress childhood memories like our parent's fights, and we repress traumatic events like accidents. The mechanism serves to keep us "in the dark" about negative experiences, qualities of our personalities we don't like, acts we commit but don't condone, and the psychological dilemmas of being alive. Often too challenging to admit, these things appear threatening to our sense of what life is about or to our concept of self. So we erase them.

We also erase awareness of negative technological events, and we forget to grasp their implications. When we turn on the lights, for instance, we forget about the local nuclear power plant that is leaking radiation or the neighborhood transformers emanating electromagnetic waves. When we paint our walls with toxic liquids and dress in clothes made of petrochemicals, we aren't thinking about the fumes we are breathing or taking in through our skin. With no further thought we become prey to, even thrive on, what Jacques Ellul calls "sociological propaganda,"[3]; we become enthralled by the latest-model cars, exercises done by Hollywood celebrities, and fashion hints, all images that are simple, unthreatening, and beg the crucial questions that lie before us.

With our minds blank, we also become prey to the "political propaganda"[4] that public relations experts consciously design to manipulate people's opinions about technology. "Should You Take a Break from the Pill?" asks the copy that looks like a scientific article but is actually an advertisement from Ortho Pharmaceuticals. The question whether a woman should even take the Pill in the first place is conveniently sidestepped. "What about the Pill and breast cancer?" the ad asks, seemingly responsibly. "Although there are conflicting reports concerning this issue," it answers, "the Centers for Disease Control reported that women who took the Pill—even for fifteen years—ran no higher risk of breast cancer than women who didn't. . . . What's more, Pill users are less likely to develop pelvic inflammatory disease and iron deficiency anemia . . . not to mention menstrual cramps."[5] Wanting the simplicity of a drug to take, we dive right into the belief system presented—and erase from our minds the 1970 headlines exposing the medical effects of artificial hormones, the women who have died, and those who have contracted thrombosis, phlebitis, and immunologic dysfunction.

In an environment characterized by repression about technology's dangers, a person who becomes ill from technological exposure finds himself alone. There is no support for or understanding of his plight. At best he will receive sympathy for being sick, but with no acknowledgment of the political implications. Even though the sick person may at one time have realized illness stems from a technological event, he may follow suit, repressing even his own knowledge. As Pat Cody reports, before she fully accepted that she was a DES mother, she used "repression and denial" to stave off the pain of awareness.

To have compassion for ourselves, we must remember that unnecessary, human-constructed disease is not easy to confront. Nor is the proliferation of dangerous technologies whose existence calls into question our dearest psychological resource: our expectations for a future. Nor is the sense of personal responsibility that arises with awareness. Faced with such facts, it is easy to blank out. Most of us do. The problem is that blanking out does not serve survival, which at this point is dependent upon our most acute awareness.

DENIAL: "It Ain't So." A second psychological mechanism we use to carry out our technology taboos is denial. This is the refusal to accept things as they are. It is distinct from repression because when we deny, we do not erase an entire situation from our minds. We merely readjust the facts to make the actual case more palatable.

Psychologist Martha Wolfenstein identifies several modes of denial.[6] One of them is disregard for an unpleasant fact: "Oh, that? It doesn't matter." Another is dissociation from feelings about it: "It doesn't bother me." A third is acknowledging an unpleasant reality while behaving as if it isn't so: "Let's go on as if it didn't happen." The most common and all-consuming reason for using denial is death. Intellectually, all of us know we will die, but few of us actually carry that awareness every day. Instead, we unconsciously disregard death's impending reality, dissociate ourselves from feelings about it, and behave as if it weren't so.

Since today's technologies have become intertwined with the possibility of causing death, it is no wonder denial of them is rampant. Denial that a technological problem even exists is quite common. Behind this stance reside defensive feelings that seek to brush aside the complexity of the issue. "The main thing about technology is that it makes life better," we say. "Technologies are just tools. They're neutral. What determines if their impact is good or bad is who controls them."

There is also denial that technology-induced illness even exists. Perhaps the most insidious form of this behavior manifest in disdain for the people suffering from diseases that are currently difficult to diagnose or prove, like chronic fatigue immune dysfunction syndrome and environmental illness. "It's all in your head," people insist. "It can't possibly be real." "You are a psycho with an atomic bomb complex," George Milne's doctors told him before he died of a brain tumor.

Finally, there is denial that modern technologies are serious threats to the survival of life on earth. "I know there is a problem," some people say, "but life will go on as it always has."

Denial often surfaces as the initial reaction to a devastating event. After the Three Mile Island nuclear accident, many officials and some citizens salved their shocked psyches with assurances that the accident wasn't really so disastrous. "Yes," they explained, "the accident happened, but it will only affect the community for a few days. Things will be back to normal soon enough." President Carter walked through the contaminated reactor on national television to show the nation that it was safe. In an attempt to downplay a dioxin fire in Binghamton, New York, Governor Hugh Carey declared he would "swallow an entire glass of PCBs"[7]—and he did.

Institutional denial of technological danger is extensive, and it has repercussions beyond the psychology of individual officials. If there are citizens who want to face the facts, without adequate information they are not afforded the opportunity. If they are in danger of becoming sick from a technological exposure, they cannot know to protect themselves. If they have already been harmed, they are not given the information they need to understand possible effects, pursue legal compensation, or take medical precautions.

Institutional denial becomes most vocal when popular concern becomes most persistent. At a time when Dalkon Shield sales were waning and three months after one of A. H. Robins's paid medical consultants reported an alarming percentage of pregnancies and life-threatening spontaneous infected abortions among Dalkon Shield users, the company put out an eight-page, multicolor magazine advertisement, unabashedly declaring the Shield's low pregnancy rates and unprecedented safety record. It was the costliest ad in company history.[8]

One of the most blatant cases of institutional denial is to be found in the history of nuclear testing. Ionizing radiation was discovered by Wilhelm Roentgen in 1896, and by 1903 researchers had realized that radiation could cause cancer. In the 1920s scientists knew it could cause genetic mutation. "By 1950," according to a lawsuit against the United States government by residents living downwind of the Nevada Test Site, "the concept that genetic alterations [could occur] at any dose of radiation went essentially undisputed among those concerned with radiological protection."[9]

Despite this knowledge, nuclear testing at the Nevada Test Site originated and proceeded with no concern for the safety of its neighbors. To begin, monitoring of fallout was almost nonexistent. What monitoring took place measured only exposures received in the first hours after a blast and avoided measurement—and therefore awareness—of exposures that took place for years afterward. In 1954, after thirty-one explosions, an Atomic Energy Commission report finally admitted that at least eleven of the tests "presented a significant hazard as to beta burn and as to whole body gamma exposure, of importance both to domestic animals and to people."[10]

Did this knowledge lead to admissions of past errors or improved safety for the future? Test site authorities immediately mounted a public relations campaign, distributing a yellow booklet called *Atomic Test Effects in the Nevada Test Site Region* to local schools, post offices, and motels. The pamphlet began: "No one inside the Nevada Test Site has been injured as a result of thirty-one test detonations. No one outside the test site in the nearby region of potential exposure has been hurt." It went on to state: "The fallout does not constitute a serious hazard to any living thing outside the test site."[11]

Refusal to acknowledge risk continues. In 1988, just as citizen, industry, and independent researchers were revealing unprecedented rates of leukemia, melanoma, lymphoma, and cancers among residents downwind of the test site,[12] the National Cancer Institute released a study announcing that there had been *no dramatic increase in leukemia or thyroid cancer among downwinders.*[13]

Recent studies of the psychology of death and dying indicate that the personal cost of living in denial can be a sense that one is not fully awake. Another cost is the sense that life itself may become unbearable, not worthy of being lived, and the result can be psychic numbing. When we numb ourselves, though, we are likely to turn to supplemental stimulation: more exciting technologies like faster cars and radical medical measures. With life that seems unlivable, we might escape into the fantasies modern technologies provide: space travel in the sky and on our video screens, robot-slaves to answer our every whim, nanomachines that promise to re-create the universe. The cycle perpetuates itself. As illness and death from technology increase, many people—from government officials and corporate technologists to average citizens—bolster themselves against the threat of what now truly becomes unlivable life. "It ain't so."

PROJECTION: "If You Were a Sports Car, You'd Be a Porsche."[14] A third psychological mechanism we use to maintain our technology taboos is projection, the act of splitting the content of consciousness into good and bad and then externalizing either those traits we deem worthy or those we deem unworthy onto persons or objects outside ourselves. A woman might complain that her colleague is manipulative when, in fact, she is being that way herself. To avoid expressing anger, a man might purposefully but unconsciously make his wife mad.

126

We project desirability onto our technologies. We think the life-style they afford us is the best. Historically speaking, we have identi-fied the kinds of technologies our society produces as expressive of our deepest creativity and our greatest power. A long-standing and seldom challenged dictum of anthropology declares that toolmak-ing—rather than ritual, language, or art—is the most crucial devel-opment to lead humans away from our animal origins. We commonly separate ourselves from and look down on "primitive" peoples, paying no heed to the fact that the differences between us and them spring not from their lack of intellect, complexity of social organization, or moral sensibility but from their lack of modern technologies.

In technological society a man driving a late-model Porsche is perceived as all-powerful. A nation fortified with stealth bombers and space shuttles becomes supreme. Cars, computers, chemicals, and missiles become reflections of potency, and with the encourage-ment of television and a commercial market that keeps cranking out new versions of old machines, we come to think of these as good, even transcendent.

A high-tech, excavated salt bed in New Mexico whose walls are cracking and floors are leaking is touted as the answer to the nation's radioactive waste storage crisis. *More* X-ray mammograms that irra-diate women's sensitive breast tissue are perceived as the solution to increased rates of breast cancer. "Star Wars" outer space defense weapons are projected to be the best way to avoid nuclear war.

The key to understanding projection as a psychological defense mechanism lies on the underside of what is held to the light. Project-ing supreme power and mastery onto our technologies reveals our deepest fear: our vulnerability before hurt, illness, and death. When we use projection this way, we come to perceive as "unworthy" the naked human, unembellished by the chrome and microchips of mod-ern invention. We see as "weak" those people made most vulnerable in a technological age: Susan Griffin, Ricardo Candelaria, Carey Wallace—the ones who have become sick.

One penalty we pay when we use projection is that we abandon our ability to observe and make distinctions. As philosopher Sam Keen says, when we use projection in the act of creating an enemy, "truth is the first sacrifice."[15] Caught by the perception that bigger, newer, and more technologies translate into empowerment, even into immortality, we lose the capacity to question their effects.

INTROJECTION: "The Experts Know Best." The flip side of projection is introjection. This is the act of taking in and identifying with everything the environment offers. We use this mechanism when we incorporate stances and standards into our psyches that are not truly our own. To psychiatrist Fritz Perls, introjection happens when "we move the boundary between ourselves and the rest of the world so far inside ourselves that there is almost nothing of us left."[16]

The attitude that modern technologies are simply wonderful is all around us—on billboards, on TV, in textbooks, on radio—and we often swallow these messages whole. I encountered a classic case of introjection behavior when I visited the public affairs director of a state agency housed in an office building that had been contaminated by PCBs. Its transformer had overheated, causing the deadly oil to vaporize and spread throughout the complex. After a year of cleanup, the National Institute for Occupational Safety and Health had declared the building safe, and I visited there a week after workers had moved back in.

It had previously been this PR director's job to facilitate relations between the agency, of which she was very proud, and the public. Now her job was to assure employees, public, and press that the building was safe. In fact, serious doubt does exist whether PCB cleanup can ever be completely effective. PCBs are chemicals used as coolants and insulators. As liquid or vapor, they are easily absorbed into the body through the skin and lungs, and they remain in body tissue indefinitely. As with radiation, which requires only one particle to cause health damage, one PCB molecule may be enough to do damage, potentially causing vomiting, fatigue, the skin disease chloracne, liver damage, or cancer.[17] According to Christopher Wilkinson of Cornell University's Institute of Comparative and Environmental Toxicology, "You can never remove every last molecule of a chemical . . . we have the capability to take care of amazingly small amounts of chemicals, but is that safe enough?"[18]

The public affairs director was quick to express doubt about the safety of PCB cleanups in other cities, and yet she was the perfect cheerleader for this one. Proudly she showed me official documents "proving" the building was safe, and she proclaimed that workers who were afraid to return were irrationally fearful.

When we use introjection, we ingest concepts and prejudices in the exact form in which they have been fed to us. As with projection, we lose the ability to discriminate or make balanced observations. If

128

we happen to take in two incompatible perspectives—"The building is safe" and "The building is not safe," or "Modern technologies are beneficial" and "Modern technologies are harming people and life"—assimilating them both becomes impossible. A person can break down in the process of trying. Since the human personality views such disintegration as the very definition of demise, it tries at all costs to avoid it. By taking on the monolithic belief that modern technologies are uniformly positive and we masterful because of them, we avoid the pain of contradiction.

PERSONAL DISCONNECTION: "The Statistics Show That 75 Percent of People Exposed Get Sick. I'm in the Other 25 Percent." A fifth mechanism people use to avoid acknowledging technology's dangers is personal disconnection. "It can't happen to me." This one is similar to denial. When we use it we admit there is a problem, but we eliminate any danger to ourselves by obscuring the facts. We explain that because of this or that element, the problem will have no impact.

In a 1978 study of United States citizens' attitudes toward nuclear weapons, psychiatrist Carol Wolman identifies a primary way people keep from thinking about nuclear war as personal withdrawal. Many participants in the study believed that only those who knew or worried about nuclear weapons could be hurt in a war. Some made statements like "I don't let nuclear weapons affect my life" and "I don't have enough of a science background to say what nuclear weapons mean to me."[19]

When I lived in San Francisco, I had a similar response to air pollution. I admitted that the pollution over the East Bay cities of Berkeley and Oakland was intolerable. I could see pink haze settling over them daily. But in San Francisco, I told myself, the ocean wind cleaned the air. Now in New Mexico, I catch myself thinking: "The air and water are clean here. The nuclear lab at Los Alamos is too far away to affect *me.*" I can see Los Alamos from my bedroom window.

Our society encourages personal disconnection from collective reality. With a high degree of fragmentation in our personal lives, with our individualized career tracks, with our schools divided into seemingly unrelated course subjects—we are not predisposed to understand that all things in life are interrelated. Several years ago my friend and colleague, Wendy, adopted a foster daughter. The girl was fifteen years old. One day Wendy said, "Let's bake a cake." She put flour, eggs, milk, and honey on the table, switched on the oven, and turned to her new daughter. The girl was looking at Wendy incredulously as if she had suddenly been catapulted into foreign territory. "What do you mean *bake a cake?*" she exclaimed. "I thought cakes came from the supermarket!"

The example may seem extreme, but in fact in our society we so dissociate ourselves from the processes of living that we often do not know what they are. The same goes for processes of illness and death. The waste dump at the semiconductor production plant leaks chemicals into the earth. They seep into the groundwater and travel through the aquifer. The aquifer feeds the wells, and well water flushes through the pipes to the faucet. Do you drink the water?

Technology survivors involved in contaminations of communities or workplaces encounter personal disconnection when they attempt to alert neighbors and fellow workers and are rejected. Despite unusual percentages of cancers, leukemias, miscarriages, and respiratory ailments among Love Canal residents, plus national media coverage of the problem, some of Lois Gibbs's neighbors refused to become involved in the issue, saying they didn't believe the chemicals could affect them. The final shock to Lois came when temporary relocation was offered and her own husband would not leave, insisting he wanted to continue working in the chemical industry.

After listening to a presentation on the health dangers of asbestos given by former shipyard worker Loran Calvert, a railroad employee in the audience confessed concern: He had worked extensively with asbestos. In an effort to help, Loran offered him membership material from Asbestos Victims of America. Forgetting one of Loran's most documented points—that one fiber of asbestos is enough to cause disease—the man jerked his hand up to ward off this information, and laughed, "Oh! *My* exposure wasn't *that* bad!"

RATIONALIZATION: "But Asbestos Is a Natural Material." Since the popularization of psychology in the 1950s, rationalization has become one of the most notorious defense mechanisms. Providing a psychological alibi, it allows us to hide the truth by cloaking real feelings and motives with reasonable-sounding explanations. The ingenious part is that the stories we make up are so convincing that we boldly tell them not only to other people, but to ourselves as well.

Just as we rationalize to explain away questionable behavior in everyday matters, so we use this mechanism to avoid awareness of the dangers of technology. "I know my smoke alarm has radioactive americium in it," we protest, "but the tiny bit in my little alarm won't hurt anyone." "Well. They're going to make computers and cars anyway, so I might as well buy them."

I have already mentioned that city health officials visiting the "sawdust neighborhood" near Ponderosa Products in Albuquerque rationalized the accumulation of dust in area homes as residents' inadequate "housecleaning." Similarly, state officials termed hedges defoliated by dioxin emitted from Love Canal "victims of winter frost," despite the fact that after a major blizzard two years before, these same bushes had flourished.

In a study conducted by Steven Kull, a psychologist at Stanford's Center for International Security and Arms Control, Pentagon officials who design nuclear policies admit, when questioned individually, that in an age of overkill, imbalances in strategic hardware are meaningless. They also admit that continuing the arms race is unsafe. In order to maintain peer identity and group cohesion, though, when these same officials meet together, they never reveal their private thoughts.[20] Instead, rationalizing an endless buildup of nuclear forces, they tell each other and the public: "The Strategic Defense Initiative is necessary for America," and "We need more efficient missiles so we can keep the competitive edge."

Whether employed by an individual, a group of people, an institution, or a society, rationalization serves to build confidence and furnish assurance about the safety and morality of one's actions. The counterfeit provided, however, does not bring an individual or society any closer to addressing pressing issues.

SELECTIVE INATTENTION: "Look at the Beautiful Skyline of the City!" Selective inattention occurs when we edit from our conscious minds unsettling elements we do in fact notice. It is an all-purpose response to daily life and probably originated as a method to filter the overabundance of stimulus we encounter. As Daniel Goleman points out, though, we discriminate when we use this filtering device, eliminating what we don't want to see or deal with.[21] My friend Marc doesn't see the dust and hair balls accumulating in the corners of the bathroom, but he is quick to notice the smudges on the refrigerator. I am a veteran vacuumer, but I never see the smudges.

Lois Gibbs describes her husband's white body stained brown by the chemicals he worked with at Goodyear Tire and Rubber. He acknowledged the fact of staining by joking about being "fashionably tan," but selectively editing out the implications, he never showed alarm, and he still works at Goodyear.

Likewise, many of us don't see the brown air pollution emanating from and engulfing the city. Federal and state officials claim they will clean up offending waste dumps, but where will they put the toxics they "clean up"? In the 1950s when fear about radioactive fallout from aboveground testing was widespread, a singular warning became popular: Don't eat snow! But who generalized the threat to include rain, streams, lakes, or the ocean?

I came across an example of selective inattention at a meeting of educators designing curricula for high school students interested in environmental action. I suggested that we consider a course to "review our collective relationship to technology." The woman recording group ideas nodded enthusiastically, turned to the blackboard, and having only heard the word "technology," wrote: "Learn to use computers."

Goleman writes that selective inattention may be the simplest and most commonly used defense mechanism in everyday life. The observation extends to include daily life in technological society. Each day we are exposed to the excesses and dangers of our technologies, many of which are quite obvious to the naked eye, yet we rarely respond to them with full attention.

The mechanisms discussed in this chapter are psychological functions that may have originated to enhance psychic comfort, cooperation, and personal and group survival. At one time they may have saved us from constant awareness of frailty or death, but in today's overtechnologized world, the smooth functioning of these mechanisms becomes problematic. In the short run they appear to promote comfort and survival, but in the long run they work against these, actually bringing us closer to death. As pesticide-poisoned Bliss Bruen expresses, "I've had visions, going down the street, of people who've lost all their hair from radiation treatments for cancer, and I've thought: We *know* what we're doing! But we're acting as if it's not an epidemic. *When are we going to wake up?!*"

At this point, our inability to look the predicament of modern technology full in the face makes it impossible for us to count on a positive future for humankind. New questions arise. By continuing to use psychological mechanisms that previously served us but no longer do, have we become the woolly mammoths of technological society? Or can we transform our awareness of ourselves and the world?

One way to broach such transformation is to ask very personal questions about our personalities: How might we receive the crucial survival information we have been avoiding—and yet maintain mental balance in the face of it? How can we reclaim our humanity and power? How can we stimulate creative, responsible thinking?

A second way to facilitate change in awareness is a more objective route: to research the technological systems our society develops by asking how they help life, how they harm it, and how we might improve, remove, or reinvent them.

A third way—and the one I offer in this book—is to befriend the people who suffer from technology-induced diseases, whether they be our neighbors, family members, or ourselves. Interestingly, in the process of becoming their allies, we encounter the opportunity to question our personal support of technology taboos, and we find ourselves challenged to rethink the social development and use of technology.

SOCIAL TABOOS

The majority . . . were encased in bulletproof ideological vests, which protected them not merely against other systems of ideas but against the direct impact of their contradictory experiences.

—Lewis Mumford, *My Works and Days*

Personal defense mechanisms explain how we in our individual minds avoid, escape, and explain away what modern technologies are doing to people's lives, but they do not explain how an entire society can engage in avoiding, escaping from, and explaining away the facts. If we view our society as a single entity, we see immediately that it exists in the grips of technology taboos too, employing what we might call social defense mechanisms. There is repression and denial when government and media neglect to generalize the daily technological disasters into a fundamental problem. We see disconnection when the Chernobyl accident becomes, in the public mind, something that happened to "them," despite common sense and evidence of health effects on both United States citizens and wildlife.[1] Then we see selective inattention: government, industry, and citizens showing more concern about the health dangers of radioactive radon, which conveniently requires no human accountability, than about those of the nuclear industry.

Because such evasions of awareness are so prevalent, we must consider them to be shaped by forces greater than merely the sum total of everybody's personal defense mechanisms. Rather they seem to be shaped by influences operating in the collective mind and governed by the fact that we live in a culture wholly enamored of and, some would say, addicted to technological development.

One such collective influence is an all-encompassing web of institutions—corporations, government agencies, the media, and the

134

military—that directly determines the content of our psychological processes. How, for instance, do we come to accept new technologies? Do we invent them ourselves? Do we understand our old technologies well enough to remake them for new tasks? Or do they "magically" appear, first on television and on billboards and then on the shelves of department stores? Does the management at work one day announce a new machine will arrive—and our lives will change because of it? How do we respond when the people next door become sick from a technological exposure? The answers to these questions are dictated only partially by our individual propensities to project, rationalize, and deny. On a deeper level, they result from the social institutions that shape how technologies are created and how we live with them.

A second influence fueling social defense mechanisms consists of the collective belief systems that determine how we think about technology. Humans will always use defense mechanisms, but the material they manipulate is not preordained. It is relative. What, for instance, is forgotten when we use denial? It is not excitement at the prospect of a shiny new machine. It is concern for the machine's impact on our loved ones, communities, and planet. What is swallowed whole when we introject? It is not the sense that human power emanates from our connection to natural forces. It is belief in technology as the ultimate source of power.

Social scientist Willis Harman writes about the power of collectively held beliefs, particularly those that unconsciously guide our thoughts and behavior. In his view a belief system stands at the core of how a society organizes itself and how its citizens view what life is all about. Its hold on the collective mind is most responsible for social blindness—or social choice. In other words, societies change when people change their minds.[2] In order to make good choices about technology, we have to become aware of what beliefs we hold. We have to realize that our perceptions, as much as we might believe in them, do not necessarily comprise all of reality. Or as Carl Porter says, "We have to learn all kinds of new things."

The social movements of recent decades have taught "all kinds of new things" to many of us. At one time many people thought women belonged only in the kitchen. At one time many people thought blacks were not as intelligent as whites and older people were dear perhaps, but useless. Many people thought poverty was the fault of poor people and Soviet and Chinese citizens were our enemies. We have changed our minds about many things.

What, then, are the beliefs and institutions that make it difficult for us to recognize that our technologies are essentially and urgently problematic? Or to acknowledge the plight of technology survivors? If we are to find our way out of the technological predicament that currently commands our destiny, we must understand the forces that create and foster it.

PROGRESS: "Progress Is Our Most Important Product."[3] In the Western world we embrace progress as if it were essential for breathing. To us, progress is the experience of time as linear, as philosopher Edward T. Hall writes, "a ribbon stretching into the future, along which one progresses,"[4] and on that ribbon we are ever forging onward, ever looking to what new things and ways we might invent.

The concept of "no limits" follows naturally. If we are, as a society, unquestioning devotees of linear progress, then we assume we can expand into infinity. We can march across land, into other people's territories, into seemingly limitless markets, into the human body, into outer space—with no attention to the health effects of the pesticides, communications towers, plutonium-charged pacemakers, plastic bottles, and space shuttles created in the service of our idea; with no attention to the infringement on the rights of people, cultures, animals and ecosystems perpetrated in the service of our belief.

The idea of progress becomes invisible and inviolable, surrounding us and informing our perceptions of human evolution and "the good life." We come to think of ourselves as occupying the pinnacle of civilization. We look ahead to the acquisition of "more and better." Industry becomes the great liberator, capable of sweeping away nationalism, militarism, and economic exploitation, leading humans to perfectability.

As a force in the creation of social organization, the idea of progress predisposes us to build institutions that can continually invent new ways and means, purvey these to ever-expanding markets, and ease the way toward further expansion. As Harriet Beinfield describes it, progress is about "how to get something we don't already have, how to make things easier, and how to make some people richer." With these goals in mind, our society creates corporations, advertising, consumerism, and governments that facilitate "progress." Large corporations get massive tax breaks to produce

136

more technologies that harm life and health, while the meager budgets of agencies like the EPA and state-level Agent Orange commissions reveal that their activities are considered worthless. "Ever onward!" is the byword of the day, and yet as David Noble says, on the level of social relations we do not change at all.[5] We do not, as a society, look over our shoulders at the human debris—the Loran Calverts and Jose Roybals—who with their lives proclaim: This idea of progress may not be so good. Writes Lewis Mumford: "Progress indeed!"[6]

TECHNICAL SOLUTIONS TO ALL PROBLEMS: "Better Things for Better Living."[7] Mumford is the modern thinker to illuminate our society's definition of progress most forcefully: It is a mechanistic, technological progress we seek, not a social or humanistic one.

"Concealed within this notion," Mumford writes,

> was the assumption that human improvement would come about more rapidly, indeed almost automatically, through devoting all our energies to the expansion of scientific knowledge and to technological invention; that traditional knowledge and experience, traditional forms and values, acted as a brake upon such expansion and invention; and that since the order embodied by the machine was the highest type of order, no brakes of any kind were desirable. . . . Only the present counted, and continual change was needed in order to prevent the present from becoming passé. . . . Progress was accordingly measured by novelty, constant change, and mechanistic difference, not by continuity and human improvement.[8]

Jacques Ellul points out that the imperative to solve all problems technically can be traced to the eighteenth century. At that time, he posits, the onset of technological society did not result from a centralized, rational conspiracy. In the wake of the breakdown of medieval society, it emanated from a shift in perception taking place in the minds of many people all over Europe. Merchants in London, shipbuilders in Lisbon, inventors in Amsterdam—all began asking what were at the time new questions: Is a particular approach effective? Is it efficient? *Does it work?*[9]

The right answers were delivered by mechanistic thinking. Captivated by scientific method and certain that reason must be applied to every facet of human life, the emerging middle class invented machines—mechanical looms, reapers, threshing machines, cannons, and rifles—and they developed machinelike social organizations: systematized monetary techniques, the factory system, bureaucratic armies, technical hierarchies that supplanted the old craft networks, and state administration by rational bureaucratic principles. In the twentieth century we still live under the systems they set up. We still look to new machines, new chemicals, and new techniques as the primary means to improve the human condition. "What does all this [technology] do for me?" asks a 1988 IBM advertisement.[10] Does it work?

Social critic Jerry Mander points out that the first perception of a new technology is invariably utopian.[11] Nuclear power will light one hundred American cities! The Pill will liberate women! Computers will revolutionize person-to-person communications! Superconductivity will make electricity as cheap as air! Genetic engineering will create a race of geniuses! Corporate-conceived advertisements deliver these messages to us via corporate-controlled media. Citizen participation in the process, from the conception of a new technology to its development and deployment, is not welcome. Besides, the process is so complex and specialized at this point that the creation of a new floor wax can be as mystifying as the development of an outer space missile system. "The result is that democratic participation in profoundly important decisions becomes impossible," says Mander, "and all decisions are placed within a technological class that benefits from a certain outcome."[12]

That class is made up of corporate, military, media, and government executives, scientists, and engineers. Many of them are the human correlates of Ellul's "political propaganda": They are the direct and conscious perpetrators of technological development. As much as the rest of us may feel we personally benefit from our cars and computers, we are in the "sociological propaganda" camp: Living in technological society, we have grown spiritually deprived. We become like the frog whose water is slowly brought to a boil. Each increasing degree of heat comes subtly enough that he becomes incapable of judging the impending danger.

Whether "political" or "sociological," people who uncritically believe in modern technology tend to salve the wounds modern technologies inflict, not by investigating the causes of technological excess, but by looking to new technologies to provide cures. Doctors treat servicemen with cancer resulting from exposure to nuclear explosions with small blasts of radiation. Chemists and farmers look to more potent pesticides to conquer the insects that have grown resistant to last year's products. Law enforcement professionals accept new criminal detection methods based on genetic examination of DNA, but do not address the social and economic sources of unlawful behavior.

A clear illustration of substituting technical solutions for systemic ones can be seen in what has happened to Native Americans in North Dakota. On the Fort Berthold Reservation, one in every three Indians has diabetes. This rate is five times the national average. Although many Indians have a genetic proclivity toward diabetes, the disease has not always been so rampant. The epidemic is technology-induced. When the United States government built the Garrison Dam, acres of Indian farmland were destroyed. The Indians lost their source of livelihood and began to live on welfare, get no exercise, and eat canned food, all factors enhancing their chance to develop diabetes. The technological response has been to send medical teams onto the reservation providing the latest treatment technologies and social work teams bringing psychological technologies to help the Indians adjust to their illness. The response has not been to explore the technological root of the epidemic.[13]

People uncritical of technology also rationalize endangering technologies by promoting humanistic uses of a particular technology. In the 1950s, for instance, nuclear weaponry was justified by its "peaceful use": cheap electricity through nuclear power. In the 1970s, when nuclear power's excesses and dangers came to light, pronuclear people tried to deflect concern by drawing attention to the medical uses of radiation.

Such rationalizations make a strong impact on both the public and the creators and disseminators of technologies. Since the notion of the technical solution has so successfully engulfed our minds, social mores, and institutions, the most seering judgment critics have been able to muster does not even question modern technology as such. Rather it asserts that technologies are neutral. They are just tools that contain no inherent political bias. If there is a problem with technology, it lies with what class of people control it.

There is another perspective. This is the technology-as-political school: Technologies serve political ends. They are invented and deployed by people who believe in and benefit from a particular political setup—and their very structure serves this setup. An overview of mass technological society shows that the kinds of technologies in place are those that serve the perpetuation of mass technological society. For instance, the telephone and computer may *look* like "people's technologies," and they do help individuals stay in communication and collect, sort, and manage information. Yet both were consciously developed to enhance systems of centralized political power. According to a manual written by early telephone entrepreneurs, the telephone was consciously disseminated to increase corporate command of information, resources, communications, and time.[14] The computer was originally developed during World War II to decode intercepted radio messages and later to boost military power through guided missilery.[15] Today these technologies make global exploitation of nature, urban centralization, and high-tech military domination not only possible, but seemingly necessary. In a decentralized, communal society, telephones and computers would be neither politically necessary nor individually attractive. As Jerry Mander sees it, "Each technology is compatible with certain political and social outcomes, and usually it has been invented by people who have some of these outcomes in mind. The idea that technology is 'neutral' is itself not neutral."[16]

The outcome of an unquestioning belief in the technical solution is to obscure our ability to see this reality. It is to invite the use of defense mechanisms that prevent us from acknowledging technology's negative effects. Most of the technologies we know and use are deployed by large institutions—corporations, government, media, the military—established to perpetuate the benefits derived by certain privileged sectors of society. Yet many people, both in and out of these sectors, still perceive modern technologies as the leading edge of human consciousness and the ultimate symbol of human perfectability. As Hiroshima survivor Kanji Kuramoto explains, "As long as the government goes on testing nuclear weapons, they won't look back to the victims they leave behind, and they won't encourage anyone else to."[17]

CONTROL THROUGH TECHNOLOGY: "Master the Possibilities."[18] Manipulation of the environment is common to all societies. Neolithic tribespeople foraged for edible roots, built fires, and painted pictures on cave walls. The Greeks fished in the Mediterra-

140

nean and constructed temples out of stone. The urge to alter one's dwelling place originates not just out of the physical need to make a livelihood, but also out of a psychological need to shape and relate to consciousness. To be conscious to the degree that humans are can be a frightening experience. We face the chaos and contradictions of our own psychology, as well as the knowledge of suffering and death.

Throughout history people have created ways—be they rituals, art, spiritual practices, or rap groups—to come to terms with the fact of consciousness. According to Jacques Ellul, technology is one of these ways. "Technique is the translation into action of man's concern to master things by means of reason," he writes, "to account for what is subconscious, make quantitative what is qualitative, make clear and precise the outlines of nature, take hold of chaos and put order into it."[19]

In the West, this tendency to put order into chaos passed from a facet of the collective personality into a dominant trait. Science was developed, in part, to control reality by creating and organizing knowledge, and technology was created in order to enact science's ideas. Today specific machines, like Dalkon Shields and jackhammers, are created to do specific jobs. Techniques, like aerobic exercises and assembly lines, are predetermined behaviors designed to accomplish specific results, and technological organizations, like multinational corporations and cities, are assemblages of machines, techniques, and people in structured relationships organized for predetermined purposes. Modern technologies exist to impose order and mastery.

At the deepest level, they reflect our Western relationship to nature: that is, both to the human psyche and to the organic world. Just as people in the West have developed techniques for controlling the mind through taboos, social mores, rules, regulations, and laws, so we have invented ways to control the external world. We have invented machines for excavating the bowels of mountains and killing insects in forests. We have crafted ways and means to enable people to murder each other and societies to make war on all of creation.

The resemblance between the kinds of technologies produced in our society and tyrannical modes of political power is too conspicuous to overlook. Langdon Winner points to the similarity in language between the two realms. *Master* and *slave* are words used to describe both the technological realm and fascist politics. So are *machine, power,* and *control.* [20] The conception, invention, development, deployment, and announcement of new technologies are accomplished by an undemocratic social process; the life experience of technology survivors testifies to this fact. But if the particular kinds of technologies in our midst exist to promote mastery and power, the question arises: Whose power? Over whom? Is the power citizens derive really in the hands of the individual word processor owner who saves paper, white-out, and time? Or is it with the corporations whose maximum use of supercomputers—performing billions of calculations each second—facilitates their mastery over vast quantities of human and natural resources? As Mander writes, the question must be more than who gets power by using certain technologies. It must be who gets *the most power?* Who is *most in control?* [21]

The premium our society places on control is so complete that it becomes difficult to see beyond it. We are encouraged to achieve mastery of our feelings, jobs, bodies, image, finances, and future. Our social institutions are busy trying to achieve mastery of disease, other nations, the shelf life of food, the seasons, and death. That the concept of mastery answers psychological needs to escape our primal fears enhances its appeal to the point that we often overlook an odd irony. The very technologies created to control life have come to express the unacknowledged, unconscious side of the modern psyche: They themselves have become uncontrollable. In so doing, they bring the fear we hoped to escape back into our hearts.

EVERYTHING, THE SAME: "We'll Be Able to Offer One-Stop, One-Shop Nuclear Weapons Simulation." [22] Stopwatch in hand, the American engineer Frank Gilbreth was an innovator of scientific management. In the fictionalized film of his home life, *Cheaper by the Dozen,* [23] Gilbreth decides that life according to traditional Victorian values is not efficient or mechanized enough for the twentieth century. His response is an experiment in time management involving not factory workers on the assembly line, but his wife and twelve children. In the experiment, Gilbreth fragments the movements of every household chore into its most essential parts and trains his children to accomplish these tasks in machinelike fashion, quick and standardized. Such an approach is what his colleague Frederick Winslow Taylor called "military organization": [24] Human movements become the levers of a machine.

The story reflects an archetype of the twentieth century. Indeed,

142

our body movements *are* determined by machines. The assembly line is the obvious example, workers moving arms and torso to accomplish the designated task as many as a thousand times an hour. In the famed 1936 film *Modern Times,*[25] Charlie Chaplin uses two wrenches to stamp indentations onto blocks flying by at breakneck speed, and when the conveyor belt finally stops, he cannot stop his body from making the same mechanistic moves.

There are less obvious examples in our everyday lives. There is the familiar thrust of the right arm shifting from second gear to third, the release of the clutch, the flip of the audiocassette into the player—actions begotten of the union of person with machine, standardized, and then stylized into "cool moves." There is the characteristic aim of the TV remote control. When we drive, traffic lights command us to stop every two blocks whether we need to or not. All these are actions we learn and repeat, like the assembly line worker, again and again and again. Just as Frank Gilbreth's children did, we mold ourselves to a narrow range of standardized motions according to the dictates of the machines around us, and when we do, we deny a host of essential physical experiences. We become numb.

Likewise, our sense of time is molded into predictable segments by the capabilities of machines. As Marshall McLuhan describes, "Electric light abolished the divisions of night and day, of inner and outer, of the subterranean and the terrestial,"[26] making it possible for people to transcend natural boundaries and stay up all night. We divide time into arbitrary weeks with people working five days and resting two, irregardless of their biological rhythms or needs. Moments become exact replicas of each other, differing only by what digit identifies them, and our life-styles, from when we eat lunch to how we make love, become a reiteration of the American Airlines slogan: "The On-Time Machine."[27]

As the painter Barnett Newman writes, human creation on the material plane is predated by the creation of an idea,[28] and the idea that came before mechanization was that life itself is mechanistic. For thousands of years cultures around the world had perceived the earth to be a living organism, a being with skin, blood, a spirit, and soul. In the eighteenth century, scientific thinking brought forth a different idea: The earth was not alive. It was dead, and nature was operably mechanistic. This idea paved the way for, among other things, detachment from and fragmentation of living, and then the complete mechanization of life.

But mechanization, one may argue, makes life easier. With it we can build houses faster and travel to all corners of the globe. The point is well made, but there are other aspects that are overlooked. Mechanization squelches the individuality and uniqueness that fed the human spirit in times past. By the fragmentation of activity and perception it demands, it makes an overview of the impact of our technologies nearly impossible. At this point, acceptance of mechanization is so thoroughly merged with cultural reality that we accept it as the way things are supposed to be. That we all use the same chemicals in our gardens—but do not know what they are made of—is accepted. That we are all exposed to the same pollutants in the air—but do not know who among our neighbors is ill because of them—becomes normal. Our vision of what is happening to our lives becomes restricted to what Langdon Winner calls a "well-trained narrowness."[29] Each sector of technological performance is so fragmented and we each, like mechanized parts in a vast machine, are so specialized in our participation that we lose sight of the whole story. As DES daughter Sarah Pirtle says, "A spell has been cast about technology, and it's everywhere. We're not even at the stage where we can question a technology when it's offered. That's a powerful spell that holds us with that much sway."

THE COMPUTER REVOLUTION: "What Exactly Can the World's Most Powerful and Expandable PC Do? Anything It Wants"[30] Half a million large computers are in use in the United States today. There are also eight million personal computers, five million programmable calculators, and millions of microprocessors built into other machines,[31] from automobiles to burglar alarms. At least thirty million Americans are avid computer operators, either in their homes or at work.[32] Production of microchips, software, and machinery is a global industry, and computerization now supports the infrastructure of the major institutions of our society, from government and military to business, media, and communications. Ever since a Harvard physics instructor named Howard Aiken unveiled the first electromechanical computer in 1944, a grand shift in both scientific and public perception of technology has been expected.

The shift is worth noting because since the introduction of the computer, beliefs about it have not, as widely predicted, supplanted already existing assumptions about technology. They have sprung from these assumptions and reinforced them. Despite all the fanfare, the computer society is still based on the same beliefs that lie at the heart of industrialism. Progress is still defined by technological development. The technical solution continues to show the way to all problem-solving. Mastery and control through technology are still thought to be the goal of human life, and efficiency through automation is still perceived as the primary mode of existence. We are still wedded to the assumptions that underlie these creeds: the mandate to functionalism, scientific detachment, and the human viewed simultaneously as device and as dominator of nature.

Seemingly new ideas ushered in with the debut of the new field of control and communications spoke of "thinking machines" that could reproduce the human nervous system. The promise of the computer was summed up by Kaiser Corporation's Don Fabun in 1967: "So simple: three things—analog, digital processing, and feedback. Put them together and you can go out in the backyard right now and out of some old bits of wire, a flashlight battery, and discarded beer cans you can . . . build a creature that can do almost everything dinosaurs did—or you, for that matter."[33]

Computers also promised infinite creativity through unlimited programming and limitless access to information—and therefore, it was thought, power and control. But these ideas were in fact based on the assumption that human psychology could be reduced to explicit, rational rules and nature to atomized elements to which such rules would apply. These ideas led to the use of computers in ways not revolutionary, but reflective of the old, mechanistic paradigm— like word processing secretaries lined up terminal to terminal like assembly line workers of the nineteenth century; like PC owners sitting alone in their rooms, separate from their families, connecting to other people only through the green glare; like government and industry officials colluding to conceal the health effects of video display terminal emissions. Instead of something entirely new, the computer became yet another attempt to express the Western notion of nature as machine.

Embedded in the zeitgeist of computerization, though—and often rationalized as its primary contribution—lies an experience that indeed does reflect a more communal and humanistic perspective than the industrial era has allowed. Anyone who has explored computer programming or tapped into a computer network can attest to this experience. It is an emphasis on overall patterns. It is the admission of the interconnectedness among all beings, things, and processes that lies at the heart of many preindustrial cultures, Eastern religions, quantum physics, holistic medicine, ecology, and feminism. People who understand the relativity of the computer "mind" see that one's perceptions are shaped by one's changing position in relation to other phenomena. The old mechanistic assumption of distinct, separate facts or phenomena is no longer relevant, and the way is paved for a more holistic perception of life. Indeed, the experience of connection among people all over the world made possible by communications satellites, personal and business computers, television, and even through the computerization of the nuclear arms race becomes a welcome—albeit terribly ironic—experience in technological society.

To apply the holistic perception computers seem to spawn, we are challenged to stretch beyond the utopian images its proponents would have us believe. The computer may offer citizens of technological society important insights that challenge some of the predominant modes of thinking, but since it springs from the assumptions that have brought us to the current crisis, it should be no surprise that it also spawns the same old package of technological shortcomings: damage to the health of electronics plant workers like Irene Baca and Susan Hernandez; miscarriages and birth defects among pregnant video display terminal workers at the *Toronto Star,* Sears, Roebuck, and other workplaces; chemical and electromagnetic pollution of the earth; and economic discrepancies between those with access to the resources amassed by computers and those without access. With it come the same old convergence of mechanistic perceptions: unwavering faith in progress, the technical solution, control through technological development, and mechanization. With it we are still riveted to the perception that functionality, scientific detachment, and human as object/dominator are "the way things are." We are not encouraged to question the total impact of modern technologies. We are still, like addicts, held in the spell.

146

THE
CARING

Even in the darkest hour, there is
that possibility of transformation.

—Wendy Grace, Saf-T-Coil IUD survivor

SELF-HELP

I don't want people to feel sorry for me. I want to stand with my head up.

—**Irene Baca,** electronics plant worker

With a better understanding of the psychic and social influences that propagate health-threatening technologies and deny their effects, let us return to the subjects of our inquiry. To sum up their experience, we can say that technology survivors often feel they have been left adrift in a world that neither understands nor cares for them. They experience victimization, loss of physical capacity and financial stability, loss of purpose, trust, and self-esteem. They find themselves faced with a host of unresolved, and often unresolvable, issues, and in the midst of so much turmoil, they often lose support from friends, family, service agencies, medical caretakers, and the institutions they deem responsible for the offending technnlogy. We do not know if certain types of technological events or technology-induced illnesses predispose a person to one or another of these claims on their psyches. We do not know if different personalities interact more intensely with one or another of them.

What we know is that the overall psychological toll can be profound. It can involve anxiety, depression, guilt, hypervigilance, social withdrawal, identity crises, and traumatic neurosis. It can lead to a downward spiral in the survivor's life by contributing to deteriorating health, marital breakup, alcohol and drug dependence, unemployment, legal difficulties, and even suicide. Carl Porter's plea—"Where's me?"—arises as the central theme.

Yet in the midst of so much psychic entanglement, another plea arises. This is the plea of the human spirit for continuity, connection, and healing—even in the most dire of situations. The impulse to transform suffering into a valuable experience has been a central theme of human myth since the beginning of consciousness. We have seen it in stories about Adam's rib, Orpheus' decapitation, and Inanna's torture. There are Job's boils, Persephone's rape, and Jesus' crucifixion. Or put in modern terms, as Pat Cody declares, "If you get a lemon, make lemonade." Gay Ducey elaborates the point. "You must work very hard not to be blighted by technology illness," she says. "You must! You cannot fail to choose. And when you choose, the experience becomes one you can grapple with and embrace." The choice? It is to move beyond victimization, acknowledge bereavement, and learn to bear ambiguity. It is to be a full human being, despite the terrible wounding.

As psychologist Michael Edelstein puts it, "Rather than merely acting in tried-and-true ways necessary to maintain the status quo, the toxic activist adopts an assertive approach to coping that involves making new relationships and finding new solutions."[1] Such an approach begins with a vital realization: "I can help myself." According to psychiatrist Henry Krystal, traumatized people can be incapable of the very resource they need: self-caring.[2] The first step is to face the fact that one needs care. It is to realize that, if only on the subtlest levels, one can provide it. As Betsy Berning says, "The change began when I realized that *I* am the person responsible for my own healing." "I want to give myself something positive," echoes Carl Porter. First and foremost, self-caring is a relationship with oneself that counters the loss of self-esteem technology-induced illness can cause. It is a relationship based not on a sense of victimization or powerlessness, but on the affirmation: "I am worth caring for."

To enter into this relationship one must vitalize a number of personal qualities. One of these is honesty about one's feelings and the willingness to express them. Over half a century ago, Sigmund Freud emphasized catharsis as a means toward disengaging one's psyche from the aftereffects of trauma.[3] In the 1970s, Elisabeth Kübler-Ross stressed "letting it all out" as an effective approach to bereavement,[4] and today's clinical treatment of post–traumatic stress disorder emphasizes the fullest possible reexperiencing of the trauma.[5]

Whether gained through common sense or psychological training, many technology survivors advocate similar approaches. As Susan Hernandez says, "The more I talk about it, the more I can live with it and the more I can accept it. It *hurts* to hold it in." Some embrace the process in a private setting they view as safe. They seek the support of other people—family members, friends, psychotherapists—who invariably ask them to recall their experiences. When Sarah Pirtle's son was in the neonatal intensive care unit clinging to life, she sought out friends who could listen to her pain. "If I didn't have enough crying time," she says now, "I would just numb out. The effect of numbing out was to abandon the baby. If I didn't cry, I couldn't stay present for him."

For others, emotional expression takes place in the public forum. Lois Gibbs describes how the endangered people of Love Canal channeled their emotions in the years before relocation. "We took our anger out on elected officials, bureaucrats, and scientists," she says. "Every time there was a release of information or a new discovery, my job was to stand there and announce it. It was usually bad news, and everyone would get mad. It was natural. If I brought in an appropriate target—the EPA, the state health department, Hooker Chemical—we had a healthy focus for our anger."

In 1980 the federal government refused, yet again, to subsidize community evacuation. Because of economic circumstances, the people of Love Canal were stuck in their contaminated homes. The EPA chose the same moment to announce the results of chromosomal studies showing genetic damage among residents. People felt trapped and terrified. A crowd of one hundred gathered in the streets and poured gasoline on a lawn spelling the letters "EPA." Then, cheering and crying, they lit the letters into flames.

In another incident a woman burst into a meeting Lois was having with a state bureaucrat. She had just learned she had contracted the immunologic disease lupus erythematosus. Her child was so upset about chemicals he refused to go to school, slept under his bed, and talked incessantly about death. "She broke into the room," Lois tells. "She started screaming, and tears were running down her cheeks. Then she picked up this enormous dictionary, like three thousand pages of everything you ever wanted to know about a word, and she *threw* it at the official!" Lois followed the woman into the hallway, and crying, hugging, and laughing, they both admitted deep relief.

Similarly, Gay Ducey's rage after discovering the source of her family's ill health was, in her words, "the most killing and cleansing rage. It was holy fire!" Concerning her son Seth, Gay describes herself as "a mother lion with limitless reserves of maternal

151

strength." She and her husband channeled the "fire" into a lengthy legal suit against the manufacturer of the herbicide that contaminated them. She believes that this process was their most effective mechanism for regaining psychological control of their lives.

"Our way of coping was to never give up," Gay explains. "We continued to battle and fight, to find resources and act on them, no matter what. Sometimes we wanted to trade lives with other people, but we didn't, and we couldn't, and we never, ever gave up. We fought corporations, school systems, doctors, and lawyers. And it was the good fight. We were fortunate to have an adversary. Not all technology survivors do. Now the court case is over. We didn't get much money, but we used the battle to channel our anger so we could move past it."

"Moving past it" is a worthy goal, yet a difficult one to attain when the disruption continues through ongoing health, financial, and legal crises. In his study of radiation-exposed soldiers, Henry Vyner describes veterans' unceasing preoccupation with health and radiation.[6] In other words, the feelings can be endless. The point is to maintain a current and honest relationship with one's psyche: to honor denial and numbing when these present the key to survival, and to express feelings when "it hurts to hold it in." The point is also to make wise choices about psychological distress.

This brings us to a second quality important for self-caring. When one's feelings are endless, the continual intensity of expression can be exhausting. At such times, a more helpful approach can be the cultivation of acceptance: that is, self-esteem in the face of painful facts that do not dissipate. Says Gay Ducey, "Being a technology survivor means you have to accept—and this was hard for me to learn—that there are some things in the world you cannot change. Now that may sound obvious, but it is not something I come to easily." Downwinder Laura Martin-Buhler also struggles with acceptance. "The people of Hiroshima and Nagasaki were able to go on with their lives and forgive," she says. "I feel I should be up to that too. The resentment and anger could pull me down and destroy me. My body hasn't let the radiation destroy me, but my mind could finish me off."

Betsy Berning has succeeded at not letting her "mind finish [her] off." "I choose not to live in regret," she says. "I don't want to do anything harmful to myself. I never know if I'll ever inhabit a body that's comfortable again, but this life is the one I've led. I have no other life to compare it to." Andy Hawkinson agrees. "I wouldn't change one second of my life. This is life, whether I'm an atomic veteran or a sanitation worker."

For some, acceptance is maximized by, as downwinder June Casey says, "counting one's blessings. I try to affirm life, not denigrate it further. Life is the most beautiful gift we have." "I try not to focus on the bad things," reveals benzene-poisoned Jane Woolf. "I try to appreciate the good things that happen."

A sense of meaning is another ingredient of effective self-caring. One of the contributions of twentieth-century psychology is the understanding that the intensity and magnitude of an experience are not the only factors determining its impact. The meaning a person places on an experience—how one interprets it, what purpose one assigns it—also contribute to how he or she reacts to it.[7]

How technology survivors interpret what has happened becomes central to their psychological well-being. The term *victim* has been used in this book to emphasize the external fact of the predicament." People going about their daily lives and with no clear choice are thrust into physical and psychological suffering because of a technological event controlled by other people. Indeed, they are the guinea pigs, canaries, and sacrificial lambs. But there is another aspect to victimization, and this is one's psychological response to it. Does one embrace the role with its attendant implications of deprivation, helplessness, and martyrdom? Or does one, despite external events, insist on dignity?

In 1982 a group of atomic veterans and other radiation-exposed people met in Fresno, California, to found an organization that would include not just veterans, but uranium miners, lab and test site workers, and downwinders as well. They called the new organization the National Association of Radiation Survivors (NARS). According to cofounder and president Dorothy Lagarreta, "At the meeting one feeling was prevalent: We're *not* victims. We're alive, and we're going to fight!" While both victims and survivors have undergone a disruptive technological event, the victim remains immobilized. The survivor draws on the experience for strength.

Genevieve Hollander's sense of strength changed through time. At first she was furious that the Dalkon Shield caused her loss of health, income, and reproductive capacity. She was angry at the doctors who did not warn her about the risks of the device and then, failing to make a correct diagnosis, prolonged the disease until she needed a radical hysterectomy. "The bottom line is that these are men who are self-proclaimed preventers of disease," she says, "and then to have them *cause* the disease and not want to take any responsibility for it—this outraged me!" Her healthy response of anger was, at first, useful, but it became chronic. "I was paralyzed for years," she explains.

Through time, though, anger alone did not define Genevieve's experience. "I realized that I'm not in control of the fact that I had this experience with the Dalkon Shield," she says. "But I am in control of how I take the event." She has come to view the world as "a stage with the evil gynecologists, inept lawyers, and suffering victims each playing their part to a tee." To Genevieve, "playing the victim can be another way of controlling the situation, of manipulating for attention, of perpetuating the whole script." Instead, she takes a new attitude: "I am responsible for every aspect of my life."

This is not responsibility in the sense of self-blame—attributing one's ill fate to negative personality characteristics like stupidity or a tendency to self-sabotage. It is not personal responsibility for hardships resulting from economic, racial, or sexual realities. It is rather *the ability to respond.* It is responsibility in the sense that: (1) "I learn from my experience"—about what to heed in the future, about my strengths and weaknesses, about my relationship to life and death; and (2) "If I am to heal, find dignity, or change my life, *I* must make it happen."

Genevieve has taken responsibility. Despite her losses and the ongoing health demands of premature menopause, she has become a successful designer. She continues to dedicate herself to healing body and psyche through diet and psychotherapy, and today she holds an attitude that gives her Dalkon Shield experience meaning and dignity. "The crummier an event is," she says, "if I can get myself to the other side of it, the better it becomes. It gives me a chance for personal triumph."

Wendy Grace has a similar attitude. She sees her illness "not as a burden, but as a gift," the ultimate context for her to grow beyond the scared woman she used to be into a more self-confident person who feels "it's okay whatever happens, whether it's healing or constant struggle or even death. . . ." To Wendy the "gift" is the meaning: a sense of personal growth she instills into her life.

For other technology survivors, a metaphysical attitude adds a sense of value to the experience. Thérèse Khalsa explains: "We are all born with a destiny to live and lessons to learn. Karma is experience you have to go through to grow. After using NutraSweet, I had a negative experience, but I learned from it." Jane Woolf agrees. "I see my life as a growing experience. Benzene poisoning happened for a definite reason. It wasn't just something that poisoned me and I was an innocent bystander. I don't believe things happen by coincidence."

Some technology survivors explain their situation in terms of religious mythology. The sacrificial lamb is the Christ who lives, suffers, and dies for a reason. "I believe in the Bible," states pesticide-poisoned Kari Pratt. "The contamination of the world fits in with God's plan, and because of this I take comfort knowing that what seems out of control to me is in God's control. The Bible says that Satan is the Prince of the Air, and so we have clouds of chemicals and radiation wafting in the wind, causing people to lose their hormones and get depressed and nasty with each other. But God has the power to overcome these evils—that is, if we follow His ways. This gives me hope. Otherwise, I couldn't take it. It's too miserable."

For other technology survivors, meaning is revealed in sociological or political terms. It is the stark revelation of the excesses of a social system. With such an attitude, the canary has a purpose. According to Anika Jans, "This society is more concerned with the survival of large, expanding institutions than with the survival of each living being. Of course I would be wounded! I'm not separate from this system. I'm in it. Sometimes I think there could be a mission in my illness. I can write and talk to people. I can help make political change. But you know? If all I am is one visible symptom of the travesty, I have not suffered in vain."

Despite the extremely wide and seemingly irreconcilable range of interpretations technology survivors give their experiences, a common element binds those that prove to be psychologically beneficial: *meaning*. Survivors imbue their experience with meaning and seek to respond to it with meaning. The experience then becomes a threshold through which they pass to achieve greater participation in life. As Laura Martin-Buhler says, "You can look at these bad things that happen as something that's going to defeat you, or your greatest challenge. If you take them as a challenge, they can be overcome."

A fourth ingredient of self-caring is humor, the uncanny human ability to step outside oneself and let go in laughter. When I asked Carl Porter what he learned from technology-induced illnesses, he blurted out: "You mean besides bad language?!" George Milne called himself the Mexican Maraca because he took so many pills to control his seizures. He told his family that he got the hiccups in the park one day and everyone started dancing!

During his interview Ricardo Candelaria related an excruciating incident. One night his leg cramped up, and he fell out of bed in convulsions. As he describes it, a nerve in his leg was bulging out from under the skin and piercing his body with pain. At one moment in the telling, Ricardo looked up at me, a twinkle in his eye, and he said: "The nerve was sticking out, *the nerve of it.*" Then, despite all, he roared in laughter.

Humor is a psychological device for expanding one's identification beyond the ego into the role of observer. It is a momentary chance to become psychically greater than one's suffering, and in so doing, to transcend it. Humor offers a way for technology survivors to revitalize a sense of connection with other people and with those aspects of themselves that are not suffering.

A last element of self-caring is the will to persevere. It is the determination, usually gained and tested by travail, to endure. Susan Griffin describes the conscious birth of her will: "At one point I saw that my physical illness and my despair were not separate. This is not to deny the physical. It's not to say that my mental condition causes my illness. That's like saying my hand causes my fingers! What I mean is body and psyche are one. You experience them all at once. Everything I do to take care of my despair helps—so at one point I made a decision. I'm *not* going to despair. I'm *not* going to slip through the hole."

Such a stance predisposes one to action. Self-caring begins with a choice. It is enhanced by honest emotional expression, acceptance, a sense of responsibility, humor, and will. It is brought to fruition by action.

The first action that many technology survivors take is to inform themselves about the nature of the offending technology and the illness they have contracted. Since the technological event that took control out of their hands occurred without sufficient knowledge, knowing becomes a means to regain what has been lost: participation in one's life.

For some the search for knowledge can be relatively easy. Information is available. After she was exposed to the pesticide chlordane in 1981, Bliss Bruen called the National Coalition Against the Misuse of Pesticides in Washington, D.C. She learned that chlordane was scheduled to be banned and that many pest control companies, like the Orkin Exterminating Company that came to her house, were quickly and furtively using up the remaining stock on their shelves. She also found out about books and studies on the health dangers of pesticides.

For those who are affected by a technology before its dangers are known, the search is a painstaking emergence out of isolation. Andy Hawkinson struggled alone in a library cubicle piecing together a picture of how, at a young age, he had contracted cataracts. He found nothing to help him understand—until he discovered and joined the National Association of Atomic Veterans, an organization that offered him the benefit of research linking eye disease with radiation exposure.

Gay and Patrick Ducey forged ahead tirelessly despite the fact that little information on dioxin was available in the mid-1970s. "I kept thinking 'Something must be done! Something must be done!' " Gay recalls. "So we did it. We became experts in the treatment of our illnesses and in the interventions for our son. We met the people at our local Agent Orange organizing committee and found them to be a remarkable source of information. The effort *to know* became a healing process in itself."

Some technology survivors become official experts. Pat Cody began as a mother, wife, and bookstore owner who read about DES in a San Francisco newspaper—and then tried to forget about it. But she couldn't. Instead, she networked with doctors and other DES-exposed people, instigated studies, conferences, and slide shows, and cofounded a national organization, DES Action. Later she became its full-time international liaison. Today Pat is an articulate and noted spokeswoman on the biology and psychology of DES exposure, as well as on political and legal developments concerning DES.

For many, self-caring action takes the form of involvement in medical care. Again, this is an act of regaining power and participation in a situation defined, from origin to repercussions, by a lack of power and participation. For people whose bodies have been harmed by medical technologies, it is particularly essential. As a result of her mother's use of DES, Perry Styles has endured two premature births, cervical dysplasia surgery, and a secondary operation resulting from that surgery. Particularly sensitive to the risks of allopathic medicine, Perry insists on active decision-making in her health care. She keeps herself informed of medical developments through her membership in DES Action. She seeks out doctors with whom she can discuss these developments, and every year she undergoes the special diagnostic tests recommended for DES daughters.

Choosing the actual mode of health care also offers the missing sense of participation in one's fate. For people whose illnesses allopathic medicine terms incurable, partially curable, or manageable only by technologically unsavory means, alternative holistic approaches provide the relief of treatments that may be equally effective in maintaining one's condition—but do not threaten negative "side effects." For some people, they provide a means to real and lasting healing.

Holistic approaches include, among others, Chinese acupuncture, herbalism, and exercise; homeopathy; naturopathy; and Ayurvedic medicine. Based on an organic, nonmechanistic perception of life, a primary principle they share is the understanding that all organs and systems of the body are interrelated. To interfere with or boost the functioning of one is to affect them all. Another common principle is diagnosis. Most holistic systems rely on subtle means like pulse reading, iridology, or symptom reporting that can reveal imbalances long before they manifest as the more grossly detectable end stage of organic disease. A third aspect they share is mode of treatment. Unlike allopathy, these systems view symptoms not as the disease itself, but rather as guides for revealing deeper imbalances that lie at the source of disease. They attempt to right the balance by enhancing each person's own capacity to heal him- or herself, and in so doing they honor that person's participation in the process of healing.

Wendy Grace feels holistic health care saved her life. In 1970 a New York doctor recommended that as a newlywed, Wendy use the Saf-T-Coil intrauterine device for birth control. Despite the IUD, she became pregnant. She had an abortion, and during the operation the doctor automatically inserted a new coil, this one larger. In 1974 new health problems began: multiple pelvic infections. Wendy also developed a recurring abscess, which the doctors treated with high doses of antibiotics continually over a six-year period, surgeries, and finally, when Wendy was too weak to endure yet another operation, by letting the abscess drain through the skin of her belly. After being extremely ill for years, she wondered if there wasn't another way.

Then a doctor discovered a tumor in Wendy's abdomen and

wanted to operate. He warned her, though, that should she experience an infectious flare-up, he would remove all her reproductive organs as well as her colon. Considering she had endured flare-ups almost continually for six years, the likelihood of this happening was high.

Wendy felt that medical science had led her down a terribly destructive road and then presented her with a dead end. She faced either the possibility of a malignant tumor and death or more debilitating treatment and downward-spiraling health.

She then learned about a holistic health center in Manhasset, New York. She mustered the courage to sever ties with her medical doctors and spent three months at the center being treated with traditional Japanese medicine, nutrition, and psychotherapy. After seven months she returned to her former doctors for a checkup. They were surprised to see that the mass had reduced to a third of its original size and subsided. Since then Wendy has continued healing with Chinese medicine, homeopathy, and physical therapy, and she is doing well.

After her experience with what she calls "the violence of medical technology," Sarah Pirtle also turned to natural medicine for healing. In 1986 a pelvic exam revealed that, like many DES daughters, Sarah had a precancerous cervical condition. Her allopathic doctor wanted to operate. Instead Sarah went to a homeopathic physician who explained that, working together, they could heal "the ground of the disease" rather than the symptoms appearing in her cervix. He prescribed a traditional homeopathic medicine, a minute dilution of a natural substance appropriate for Sarah's condition. He also gave her a distillation of diethylstilbestrol (DES) whose purpose was to reverse the process of pathology at its root. The cells of Sarah's cervix turned pink with vitality.

Marie Ferneau's story is notable because the centerpiece of her healing process was her own intuition. If a major criticism of allopathic medicine is the all-powerful authority of the physician whose judgment and technology may cause harm, a principle of holistic health care is to return knowledge and participation to the patient. Marie reclaimed her authority in this relationship to an unusual degree. She used medical doctors for their expertise at diagnosis, and in the face of no recommended treatment from them, she used her intuition to guide her own health care program.

As a sculptor in the 1960s in Tulsa, Marie attended an industrial lecture on sewer pipe casting and learned about a new material for casting previously uncastable shapes. The material was Styrofoam. Information about the health dangers of industrial materials was virtually unavailable at the time, and she set about melting Styrofoam and using other toxics like water glass, fiberglass, acrylics, and automotive chemicals. Within a few years Marie went blind in one eye and suffered recurring viral infections. A battery of medical tests revealed that she had a degenerative condition made up of a near-complete blood chemical breakdown, hypothyroidism, adrenal collapse, osteoporosis, and seared lungs. An opthalmologist she saw in Phoenix proclaimed, "Marie! You *can't* be alive!" But aside from recommending surgery to replace her lenses (which she did), doctors had no restorative program to recommend.

Feeling resigned, Marie decided to try to adjust to her condition, and she went to a psychologist. Something astounding occurred. Under hypnosis, she discovered that she had an intuitive ability to gain otherwise unattainable knowledge. First, while in a deep trance, Marie saw the source of her illness. In her mind's eye a piece of Styrofoam and the letters CY appeared. She later telephoned a local manufacturer of the product and learned that among other chemicals, Styrofoam contains cyanide. Marie had been breathing cyanide fumes.

She then embarked upon a meditation practice asking her unconscious mind each day: "What can I do today to help?" "My discovery of intuitive power," she explains, "plus the mental breakthrough that I could heal myself enabled me to focus all my attention away from the disease and onto wellness." One morning Marie would perceive a glass of lemon juice, and that day she would drink lemon juice. One morning she would see her feet soaking in hot and cold tubs, and she would do this. A constant in Marie's intuitive health program was to take hot baths and saunas. Another was to drink a half cup of olive oil each day. A third was, via images of truckloads of carrots, Chinese cabbage, avocados, and watermelons, to ingest megadoses of vitamins A, C, and E, and potassium.

By the mid-1970s, Marie had recovered her health. Although her lungs remain sensitive and eyesight limited, subsequent blood tests show a reversal in the degenerative process with a restoration of adrenal and thyroid functioning. Today Marie's blood chemistry is normal, her bone mass is normal, and she works as a meditation counselor and writer.

It is interesting to note that in the late 1970s, scientific research produced the Hubbard Method for reducing body burdens of various chemicals. L. Ron Hubbard originally developed his program to detoxify psychoactive chemicals from addicts at drug rehabilitation centers,[8] and other researchers have proven it up to 66 percent effective in reducing body burdens of PCBs, PBBs, and some pesticides.[9] Interestingly, the program includes: (1) forced sweating by sauna to mobilize chemicals out of the fat tissues where they are stored; (2) ingesting polyunsaturated oil to scavenge chemicals from the blood for excretion; and (3) ingesting vitamins and minerals to avert dehydration and replace those lost in sweating.[10] Perhaps most compelling is the fact that the Hubbard Method was not publicly reported until 1980, several years after Marie had executed a self-help program identical to it.

Another positive action technology survivors take is to create social support systems for themselves. Since networks based on ties to family, friends, and coworkers and support from government institutions too often fail, survivors find they must improvise alternatives. As Asbestos Victims of America executive Heather Maurer explains, "We have no one else to turn to except each other."

The original impetus for forming self-help groups is the plethora of problems survivors face, problems they share with no one else. These problems often form the core of a common identity that, as Michael Edelstein points out, becomes more important to the survivor than normal categories of identification like politics, social groupings, or geography.[11]

After contamination at the GTE plant, Susan Hernandez and Irene Baca joined with over one hundred coworkers. One purpose of the association was to pursue a legal case against the corporation and the chemical manufacturers. Another was personal support. "We're holding each other's hands," says Susan. "We're all in the same boat." According to Gilberto Quintana, "When I went to the NARS conference in 1986, I learned there are other people who have the same troubles as me. Everyone there knew what I had been through."

Some self-help groups are founded to link people who share injury by a common technology, and these are often national or international in scope. Examples include Asbestos Victims of America, International Alliance of Atomic Veterans, Aspartame Victims and Their Friends, Dalkon Shield Information Network, and Citizen's Clearinghouse for Hazardous Wastes. Others address a local technological event that has influenced a geographic community, such as Silicon Valley Toxics Coalition, Northwest Coalition for Alternatives to Pesticides, Heights Area Residents Against Pollution, Warwick Against Radium Dumping, Los Amigos del Rio, and Love Canal Homeowners Association.[12] June Casey is making an effort to reunite with her sorority sisters from Whitman College, all of whom are radiation survivors, whether they know it or not, from the 1949 Hanford release.

A benefit of creating self-help groups is the chance to share information, set up resource files, and pioneer research. Another is to revive some lost psychological strengths: a sense of involvement and self-esteem. According to Martha Wolfenstein, the sense of abandonment that follows catastrophic experiences can be ameliorated by the discovery that one has not been singled out for misfortune: There are others who understand.[13]

Marmika Paskiewicz was exposed to chemicals like xylene and 1,1,1-trichloroethane at her job as a museum exhibition specialist. After organizing coworkers to share fears and medical symptoms, she reported: "We had to work to strengthen our self-confidence so we could say out loud, 'This is killing us, and we're not going to do it anymore!'" According to Carl Porter, his self-help group was a "literal lifesaver." Telling his story, hearing others', and sharing his feelings helped him to "accept my illness, make some real friends, and rediscover a long-lost ability to laugh."

Whether technology survivors help themselves by acknowledging feelings, cultivating psychic resources, making health care decisions, or starting a self-help group, the long-range effect of making the choice to care contributes to both physical and psychological healing. The immediate effect is, as Gay Ducey asserts, to make the experience one "you can grapple with and embrace."

GOD'S HELP

I found Him up there.

—Gilberto Quintana, atomic veteran

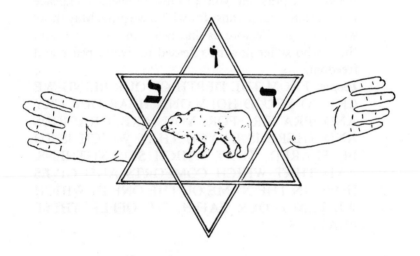

—Bear Kornfeld, drawing for bar mitzvah
invitation featuring Star of David,
totem animal, and his own four-fingered hands

Lector: The holy presence in our midst who stands with
those who suffer, who charges us to be the voice of the
voiceless: We call upon you for all our brothers and
sisters, mothers and fathers, sons and daughters who
have suffered and died of radiation exposure. Bless their
souls and comfort their spirits. Let not their deaths be
in vain, O Holy One, but create upon their sacred sacri-
fices a new vision and a new world that is safe, free,
peaceful, and just.

People: FROM THE DEPTH OF OUR BEING WE CRY TO YOU, O HOLY ONE!

Lector: For those who have suffered and died in Hiroshima and Nagasaki, Bikini and Eniwetok, Kwajelain and Mururoa, Fangataufa and Christmas Island, Johnston Island and Monte Bello, Emu and Maralinga; for those whose land and sea are today being put at risk through radioactive pollution, from the dumping of nuclear wastes, and the passage of nuclear ships.

People: FROM THE DEPTH OF OUR BEING WE PRAY TO YOU, O HOLY ONE!

Lector: We pray that your promise of justice and peace may become real to those for whom we pray. May those who have gone before us find peace in your sight and those who suffer now be released to live in peace and freedom.

People: FROM THE DEPTH OF OUR BEING WE CRY TO YOU, O HOLY ONE. HEAR OUR CRY AND PRAYER, FOR YOU ARE GRACIOUS AND THERE IS IN YOU THAT WHICH IS TO BE FEARED, THAT WHICH STRENGTHENS, AND THAT WHICH COMFORTS AND GIVES HOPE. IN THE NAME OF THE ONE IN WHICH WE PLACE OUR FAITH, WE OFFER THESE PRAYERS.

—**Reverend Nobuaki Hanaoka,** litany from memorial service for Radiation Survivors Congress, October 1984

A scene in the ABC TV movie about nuclear war, *The Day After,* shows the desperate townspeople of Lawrence, Kansas—clothes tattered, ashen, and withering with radiation sickness—listening to a minister preach about God. The scene reveals neither God's equanimity nor the vigor of the people's faith. It is a pathetic commentary about absurdity. Indeed, as Robert McIntyre and Kari Pratt report, many fellow residents of pesticide-poisoned Fort Davis, Texas, feel disenchanted with "a God who would allow this." Similarly, Henry

Vyner tells of an atomic veteran who gave up his faith in religion after he saw the blast, saying, "If God is *that* angry to unleash that power, then Christianity no longer works."[1] For other technology survivors, though, usually those who experience God not as a personification of "otherness" but as a living power within, faith is miraculously bolstered.

Throughout history religious mythologies about Jesus, Persephone, Abraham, and other archetypal figures have provided an inroad to universal questions about the meaning of life. They have given people a framework for interpreting human psychology and guidance toward a positive experience of living. But beyond the stories that characterize all religions, what is God? How does spirituality help a suffering person? How does one attain it? The people who grapple most directly and starkly with questions of life and death are often the ones most prepared to answer these questions.

According to Dalkon Shield user Betsy Berning, "What opened me up was the *dead end.*" One view of spiritual awareness perceives the loss of identity and opportunity resulting from illness as a threshold to inner growth and spiritual experience. This is the "sacred wound" theory. According to psychologist Jean Houston, the process she calls soulmaking often begins with "the wounding of the psyche . . . a painful excursion into pathos, wherein the anguish is enormous and the suffering cracks the boundaries of what you thought you could bear . . . [requiring] that you die to one story to be born into a larger one."[2]

For most people the old "story" is sociologically defined. It concerns one's meaning in life in terms of family, achievements, possessions, and other external phenomena. As Teddy Ostrow says, "I was very concerned with my image and my job and amassing tokens of success. I probably would have gone on living like that . . . if the exposure hadn't happened." A "larger story" concerns one's relationship to universal patterns and to the essential force of life. Through the wounding initiated by a technological event— through the breakage of one's body, identity, trust in society, and sense of the future—one's normal expectations are extinguished, and, as Houston writes, "the violation of these boundaries makes us vulnerable to be reached by larger forces, by the Larger Story."[3]

Another view of spiritual opening is psychoanalytic. It pinpoints a predictable pendulum between the wrenching feeling of abandonment so often associated with illness or catastrophe and the secure and wanted feeling of union. According to Freud, the person who endures a catastrophic event can feel abandoned, become angry, and lose faith, but then in time experience a reintensification of faith equal to the intensity of the disillusionment.[4] Martha Wolfenstein emphasizes this dynamic in her work on disaster. The abandonment/union cycle occurs not just in social relations when people who have endured a disaster band together in community which, for many of them, provides the most rewarding sense of kinship of their lives. It also occurs in people's relationship with, as Wolfenstein writes, "the superhuman." After a tornado or an air raid, survivors' feelings of terror and separation are often followed by a profound awakening to forces greater than themselves.[5] Likewise, after technological events like toxic exposures, IUD episodes, and nuclear blasts, survivors often look to or undergo spontaneous reunions with God.

The two views differ in their philosophy of spirituality. The "sacred wound" approach sees it as a real and essential aspect of consciousness, one that exists whether or not a human is aware of it. The psychoanalytic perceives spirituality as a human-centered, psychologically induced need. The debate raised by these opposing views cuts to such a personal level of interpretation that its resolution may never become objective. What the two views share, though, is the observation that attunement to spiritual experience can have the effect of healing and redeeming the sufferer—and this is what is most important to the person who needs healing and redemption.

What is spirituality? It is enhanced by certain attitudes, by postures one strikes in approaching life, such as Jane Woolf's perspective that "life is a growing experience" or Wendy Grace's notion of "illness as a gift." Yet in itself, attitude does not constitute spirituality. It is merely a door that may help one find it. At its root spirituality is an experience, a way of being. It is the direct experience of and identification with the energy that is universal to all life. And it is difficult to verbalize. Interviewees, asked what resources they had found or invented for coping, answered "God," "the soul," "spirit," "Goddess," and "the Big Picture."

For downwinder June Casey, this is a feeling of "abundant sustenance"; for atomic veteran Ricardo Candelaria, "relief"; for pesticide-poisoned Kari Pratt, "peace and comfort"; for atomic veteran Jose Luis Roybal, "faith in God is what keeps me going. It is something stronger to keep on holding you." Robert McIntyre explains the experience as one of "power." "God reassures me," he says.

166

"When things are the bleakest, I have someone to go to. Whatever happens to me, I feel assured that I'll be taken care of. I don't completely understand *why* I have this illness. The question continues to be a mystery to me, but it's not a mystery sufficient to keep me from honoring God. He's a tremendous resource for me. Gosh, there's just *such a power there!*"

When people discover this power, they often choose a religious form to encourage and express it. They become Christians, Buddhists, or Jews. They practice meditation or pursue a teacher. One of the chief characteristics of the state of mind, though, is a fluidity, an openness to the moment, a lack of dogmatism. Ultimately, the experience is one of connectedness beyond human categorization.

Laura Martin-Buhler, who is a Mormon, feels that prayer has been her most effective way to attain this sense of connection. "When I pray," she explains, "the first order of business is to thank God for my blessings, to be grateful. Then I ask for what I need, whether that's money, medicine, or a miracle. But what really helps is more than thanking and asking. I don't pray by rote or jump up from my knees when I'm done. I open myself to listen. Meditation is the most important part of prayer. It's like being a radio. I have to be tuned to the right channel to hear the music."

The detachment implicit in this frame of mind is not detachment in the sense of "I don't care." It is a witnessing that involves the presence both to experience one's feelings fully and to transcend them at the same time. The challenge is to be involved without getting entangled. Wendy Grace explains: "I've learned there is a difference between my personality and the soul. The soul is universal and bigger than my person. You see, at the moment I am lowest, there's a humility, a giving away of my personality's preoccupations with getting out of the situation, or getting what I want done, or being afraid. That moment is a moment of power. It's an inner dawn. The world changes a bit. On the deepest level, I don't see myself as sick. I see myself swimming through layers of my personality to a sacred world. . . . Recently my swimming has shown me that my suffering is not personal. I'm part of the human species, and this world of pesticides and bombs and pain is where we are now. My personal struggle becomes a struggle for the world, and when I know this, something in me gets bigger. I don't feel so alone. I feel powerful."

For many technology survivors, developing the ability to observe and yet be present for life's pains and joys has required that they learn to trust their own sense of things. The challenge can be difficult in a society that so thoroughly honors knowledge garnered by scientific method and denigrates that obtained subjectively. "I learned to trust my own inner voice," reports Betsy Berning. "I've slowed down," says Thérèse Khalsa, "so now I have time and presence enough to find out what I know deep down." Harriet Beinfield concurs. "Ultimately, there's nobody who knows what's going on better than you do."

Another quality important to spiritual connection is a sense of faith. This is not blind faith that "things will turn out." After enduring the many breaches of faith technology-induced illness can initiate, most survivors become too apprehensive to wish that problems will disappear or life return to normal. This faith is rather an active presence of mind. It is an experience of fullness in the here and now, enhanced by dismissing the doubts and cynical thoughts that can extinguish what joy and meaning are possible. Anika Jans describes it as "living in faith," an act bolstered by one's ability to feel connected to forces greater than oneself. Wendy Grace describes it as "a state of grace, a love deeper than I have ever known, something beyond my attachment to what does or doesn't happen. It's peace. It's my lifeline."

The benefits of this expanded sense of identification can be very practical. When bad news threatens, one does not immediately sink into despair or resignation. One can expand awareness to a sense of reality and possibility beyond the physical. If depression cannot be avoided, particularly if illness biologically induces it, one can seek identification with the greater realm, and while depression may still impinge, it does not command all of a person's attention. It exists in a larger context. As Jose Luis Roybal so simply puts it, "Have faith. Even if things get worse and worse, *it's a better way to live.* Have faith."

Having faith includes knowing love. There is compassion for oneself. This is crucial to the technology survivor whose inner life may be so filled with turmoil and negativity that she has lost all sight of self-respect. "If I can't feel well," says Anika Jans, "my life isn't the way I want it to be. But I don't have to beat myself up. I'm here. I'm okay, despite everything. Who I am, ultimately, is something much greater than my earthly desires."

The expansiveness of compassion flows to other people as well, and it offers the chance to eradicate that brand of alienation and isolation that originates in one's own mind. Gilberto Quintana goes into the hills of New Mexico each Lent to pray, and when he returns, he offers what he has learned to his fellow townspeople through spiritual counseling and communion.

"I've learned how to love," says Andy Hawkinson. "I've learned how to recognize those who know how to love and feel and share and have compassion. It's something that emanates from a person that says 'This is a person I can be close to.' If nothing else, I've become a better individual. I've loved more because of my illness." "As soon as I saw that Bear had been born with no thumbs," says Harriet Beinfield, "I became aware of millions of people, like the parents of children who never get better."

To the psychologist F. Haronian, a relationship of empathy between people springs from and enhances the spiritual experience.

> To meet at the interface of I and Thou means that each member . . . is willing and consciously as close to the other as possible; so close that for each person the subject/object dichotomy is blurred and more or less dissolved. Ego needs are not operative. At this point, one self meets the other, and those disinterested aspects of each individual merge for a moment in a sense of union.[6]

This experience of connection—whether attained spontaneously or with great effort, whether manifested directly to "the superhuman" or through other living beings—invariably brings one to a deeper appreciation of being alive. "I've learned how precious life is," says Wendy Grace. "If we give ourselves to God," says Jose Luis Roybal, "we can thank Him for every day." Reports Andy Hawkinson: "I've learned to cherish the good times and never forget the times that are bad."

This deeply felt appreciation for life, in all its pain and glory, with all its suffering and joy, can lead to an acute sense of celebration. Gilberto Quintana contends that people who "do everything one day at a time" bring more enjoyment to moments of celebration. Gilberto certainly brought all of his vitality to the 1986 NARS convention party, the Crossroads Ball. There with hundreds of fellow radiation-exposed people, amid silver and blue helium balloons, sixty-year old Gilberto took off his hand-painted "Atomic Veteran" jacket and jitterbugged, two-stepped, and waltzed all night. It was a joyous time for the atomic widows who had lost their husbands to leukemia and cancer, dancing for hours with *hibakusha,* lab workers, downwinders, and veterans.

Sometimes celebration is expressed in the simple joy of being alive, sometimes in the joy for a victory. It can also be expressed in very private moments, as it was in heartfelt hugs exchanged by GTE workers' counsel Josephine Rohr and her clients when they won an out-of-court settlement against the corporation in 1987.[7] The day the federal government announced its intention to pay for the evacuation of Love Canal residents was a day of sheer ecstasy. As Lois Gibbs recounts, "Someone brought a case of champagne. Corks were popping. We were so happy. People were laughing, crying, hugging each other, dancing around, and saying 'We won! We won! We're out!' Our babies would be safe from further exposure to Love Canal poisons!"[8]

The feelings of glory, sustenance, power, and celebration fly in direct contradiction to the grief, lowered self-esteem, guilt, and fear that are common psychic responses of technology survivors. To find a way to both accept and transcend these difficult states is in itself a miracle. The survivors I interviewed who have developed ways to attain spiritual awareness report greater psychological stability in their lives than when they did not have access to a spiritual perspective. Most of all, they report a renewed sense of trust in life, as Wendy Grace says, "a state of grace."

170

HELPING OTHERS

If we can't find a cure for me, I hope I last long enough
to make it possible for someone else to be helped.[1]

—**Bob Waters,** professional football player exposed
 to synthetic fertilizer on playing field

God left me here for one purpose: to help other people.

—**Gilberto Quintana,** atomic veteran

The compassion that often swells in a technology survivor's heart
can flow into their hands, becoming a desire to help other people. In
her work on disaster survivors, Martha Wolfenstein explains that the
impulse to generosity is extremely common. At first, feelings of
danger catalyze people to focus exclusively and defensively on their
own survival needs: "Help me!" A polar reaction follows, an expansion toward concern for others in need: "Can I help you?"[2]

This is a psychic turnabout that allows the survivor to challenge
the helplessness and isolation that can predominate by refocusing on
empowerment and participation in life.[3] Downwinder Laura Martin-
Buhler explains: "Anyone who has suffered has more empathy for
others who are suffering. I try to transform my own painful experiences to the good of others. Service is the best antidote for personal
suffering. It gets you out there, and you try to help other people get
to the place where you are trying to be."

As Carl Porter makes clear in his description of the self-help
support group that brought other people's struggles into his awareness, caring for others can reduce a survivor's anxiety by focusing
attention outside oneself. "When you're involved in the suffering of
other people," explains June Casey, "it's much easier to accept your
own."

Helping others can also provide meaning to a life from which purpose has been torn. "I feel blessed to have made it through, and I know there is a reason for being alive," says Laura Martin-Buhler. "Survivors of any catastrophe acquire a sense of destiny. Life is not to be taken for granted, and God has left you for a purpose: to bring something positive to other people."

Lastly, caring for others offers the chance to expand one's identity beyond that of "sick person" and "maligned victim" into one of "helping hand" and "ally." The act of giving can also be such a powerful experience that it challenges momentary attachment to any identity at all, leaving one open to a sense of spiritual connectedness that can be more rewarding than the provisions of ego identity.

Perhaps the most direct channel for technology survivors' impulse to generosity is presented in cases of community contamination: The negative technological event has affected everyone in the immediate vicinity, and everyone needs help. In places like Times Beach, Missouri; Woburn, Massachusetts; and Kanab, Utah, ties to family, friends, and local institutions are directly available—right next door and down the street. All that remains for the survivor is the invention of how to serve—yet because so many survivors feel isolated in their predicament, even when in community, reaching out to help fellow human beings demands more than conventional, pre-patterned reactions. It involves acts of creativity outside the common repertoire of behavior and beyond the common boundaries of thought.

Like many people, Robert McIntyre moved to Fort Davis, Texas, because it was purported to be one of the most environmentally clean areas in the United States. Having suffered poisoning from the mercury in his amalgam dental fillings, he looked forward to living in Texas and healing from paralyzing allergies, chronic fatigue immune dysfunction syndrome, and candida infection. He also looked forward to reestablishing himself in his profession as an attorney. Likewise, Kari Pratt came to Fort Davis because her immune system had deteriorated in Iowa after exposure to the pesticide lindane. After four years in Texas, she was progressing well, looking rosy-cheeked, and was "almost normal."

Then in 1987, at least six vineyards nearby began to spray paraquat, captan, Bayleton, Sevin, and other pesticides on their newly mature grapes. The poisons drifted in the wind throughout the region, and Robert and Kari reacted immediately. Each became ex-

172

hausted, and their sensitivities to foods, pollens, and chemicals, so newly healed, returned. Today every time a burst of pesticide toxin wafts by on the wind—and it did several times during our interview—Robert detects a metallic taste in his mouth, and both he and Kari became dizzy and depressed. And when the spraying becomes so overpowering in Fort Davis that living there becomes impossible, Robert offers his truck to Kari, and like fugitives in a poisoned world, the two of them drive, sometimes all night, to escape the clouds of toxins. When I met them in July 1988, they were on the run from a recent spraying and wouldn't be able to return home until the vapors would subside—in October.

The urge both Kari and Robert feel to save themselves is complemented by a passion to help the other people of Fort Davis. Doing door-to-door research in the community, they found that not only were environmentally sensitive people in the area suffering, but since the spraying had begun, non-environmentally ill neighbors had become sick as well. As of July 1989, Kari, Robert, and staff members at the Emergency Medical Services estimated that some three hundred people in a community of twelve hundred were suffering from new cases of stress hypertension, high blood pressure, high cholesterol, asthma, arsenic poisoning, Bell's palsy, malignant tumors, and miscarriages. They also estimated that 80 percent of these cases had struck people who were not environmentally ill before. Their response to the situation has been to speak with these people, inform them of the possible link between pesticides and encroaching illness, and educate them about how to care for themselves.

Perhaps the most compelling aspect of Robert's and Kari's desire to help other people, though, is enacted in their relationship with one another. Having both struggled with debilitating illness before the spraying began, they each have empathy for the other's suffering. Since it has begun, they have become best friends. They help each other by doing things together like fleeing the poison pesticides, by doing things for each other like buying each other vitamins and medicines, and by listening to each other's sorrows and hopes. These ways of caring are typical. Among technology survivors a distinct desire to serve is evident, and they use, find, or invent all kinds of ways to express the urge.

They help people by joining them in common activity, what psychologists Marc Pilisuk and Susan Parks call "social exchanges"[4] and psychologist Shepherd Bliss calls "side-by-side" caring.[5] Since he could not have children of his own, atomic veteran Ricardo Candelaria rounded up the children of his neighborhood and diverted their minds away from crime and drugs by starting a baseball team. He also loves to pass down Hispanic culture by regaling the children with stories about how life in the barrio used to be. Gay Ducey spends extra time playing with her son Seth and practicing the special skills he will need in life as a disabled person. DES mother Pat Cody accompanies her daughter to important DES medical examinations and joins her in hashing out the decisions about health care that DES daughters must consider.

Technology survivors also help by doing tasks for other people. Some help their families. Carl Porter felt elated to help his eighty-year old parents after their house burned down. Worried that exposure to a plethora of toxic chemicals in new building materials would "finish them off," he researched nontoxic building methods for their new house, saying "They're joyous, happy people who could go on another twenty years—if they're not exposed."

According to former asbestos worker Nathan Robinson, winning a settlement from an asbestos manufacturer "helped me by letting me help someone else." He bought a house for his daughter. Robinson also derives satisfaction from working in his community, volunteering for the Lions Club project to provide guide dogs for the blind. "I get a big bang out of helping people like that. It makes my day. It makes my week!"

Gilberto Quintana works in his northern New Mexico community running bingo games for the Disabled American Veterans, serving on the board of directors of the senior citizen center, and cutting hair for people who can't afford professional care. "Senior citizens. Sick people," he says. "I am always doing something for somebody. That's what has me going. And I'm going to talk to the history classes here in Pecos. I'm going to tell them to be really careful. When you are out there in the work force, you don't know what you are going into. Neither the company, the government, nor the foreman gives a damn about you. If you don't watch out, you're going to land the same thing like me."

As Gilberto emphasizes, sharing knowledge is an important way technology survivors help other people. Support groups that begin for self-help for a small number of people often evolve into organizations that reach out to help others. The Napa-Solano chapter of Asbestos Victims of America (AVA) is a bustling center in Vallejo,

California, with former asbestos workers sharing a Victorian house with another service organization, the Federation of Employed Latin American Descendants. There, chapter coordinator Jesus Rives spends his days giving information to the many people who call for help. He helps a bedridden asbestos worker find a lawyer with a walk-up wheelchair ramp. He helps an asbestos widow prepare for a court case and a former shipyard machinist to understand his exposure. National board member Loran Calvert sits at another desk setting up public education forums, encouraging the asbestos-exposed to join the organization, and "helping others go on what I think is the right track." Says Jesus: "For asbestos victims we're the heart and we're the arms. Helping people. There's no other way."

DES Action also began as a group of concerned people looking to help themselves and grew into a national organization serving the DES-exposed population around the world. Pat Cody was one of its founders. Starting with fear and little information, she located other DES mothers, networked with medical professionals, and did research. Concerned that the majority of DES sons and daughters weren't aware of their predicament, Pat participated in producing the first leaflet ever written on DES exposure. She helped launch a two-year consumer education program, develop a slide show for clinics called "Ask Your Mother: Finding the DES-Exposed," and advise the producers of *The Lou Grant Show* for a program on DES. She also spoke about DES at the 1985 United Nations Conference on Women. Her main motivation is to help: help people who want children to have them, help those with cancer to get care in time, and help those with health effects to distinguish between medical treatments that can exacerbate their problems and those that can help.

Sarah Pirtle joined DES Action not just to inform herself, but also to inform others. After what she calls "a disastrous experience of hospital negligence" at Ryan's birth, she wrote a booklet for mothers of premature babies and an article for the *DES Action Newsletter* "so other women wouldn't have to go through the same thing." Likewise, Perry Styles set up "DES Awareness Days" and was awarded a state grant to show films, give talks, and make contact with physicians at seven Massachusetts colleges.

Organizations like AVA and DES Action also help other people by lobbying for legislation to benefit their constituents. The Committee of Atomic Bomb Survivors in the United States does too. This organization is made up of Americans of Japanese ancestry who were living in Hiroshima and Nagasaki in 1945. Since returning to the United States after the war, they have faced the many medical, emotional, and financial hardships of radiation-exposed people—with no aid from their government. Since 1971, members of the organization have steadfastly worked to help all American survivors of Hiroshima and Nagasaki by lobbying elected officials for a bill to provide medical treatment.

Both the National Association of Atomic Veterans and the National Association of Radiation Survivors also spend much of their resources lobbying government officials for legislation to help radiation-exposed people—and both were instrumental in the passage of the 1988 Radiation-Exposed Veterans Compensation Act designating D.V.A. service-connected disability and death benefits for thirteen cancers.

Citizen's Clearinghouse for Hazardous Wastes in Arlington, Virginia, was inspired by Lois Gibbs's experience at Love Canal. She started the organization to help people facing contamination by toxic waste. Today she and her colleagues advise thousands of people all over the United States in their struggles to save their communities from toxic incinerators, landfills, and dump sites. The headline on their brochure reads: PEOPLE HELPING PEOPLE TO HELP THEMSELVES.

Other technology survivors feel that their legal cases against perpetrators of dangerous technologies provide a means for helping other people. Concerned that aspartame is still available in food stores and restaurants, Thérèse Khalsa says, "I would sue the NutraSweet company not for myself, but if it would protect other people from going through what I went through." Before her lawsuit against GTE, Susan Hernandez looked forward to testifying so that "other people could think about their employers' responsibility for safety." Gay Ducey explains her court case against a major oil corporation as a means to make the company monitor its endangering products in the future.

For other technology survivors, helping people by sitting with them and listening to their feelings is most satisfying. Gilberto Quintana brings both the depth of his feeling and his detachment when he counsels alcoholics in his community and sits with fellow atomic veterans sick with cancer. Laura Martin-Buhler volunteers her time at the Esperanza Center for Abused Women in Santa Fe, New Mexico. Andy Hawkinson has traveled around the United States forming rap groups and listening to veterans express the suffering induced by radiation exposure.

For some the personal experience of technology-induced illness is such a dramatic turning point that they dedicate their lives to caring. They become healers. Wendy Grace explains that she became a massage therapist as a result of attempting to heal her own illness. Receiving various physical therapy treatments to nurture and balance her own body, Wendy realized she wanted to work in alternative health care. She studied anatomy, massage therapy, Japanese acupressure, and the Rosen Method, an approach akin to laying on of hands. Today she has a private practice revitalizing her clients' capacity for self-healing by placing her hands on their bodies and "channeling energy" to them.

As with all acts of giving, Wendy finds that the work is healing not only for the people who come to see her, but for herself as well. "The fact that I'm doing body work constantly contributes to my own health," she says. "The energy goes through me to the person I'm massaging. Each time I work I have to process whatever energy passes through me. It's like doing tai chi exercises or giving myself an acupuncture treatment. If the flow stagnates in my system, I have to clear it out. The work challenges me daily to self-awareness and self-monitoring."

Laura Martin-Buhler also became a massage therapist as a result of her technology experience. She is such a gifted healer that despite the fact that she never took classes, a nationally renowned massage academy awarded her an honorary degree. Laura's philosophy: "My greatest desire is to make other people happy."

Likewise, Harriet Beinfield brings a world of experience to her profession as an acupuncturist. Citing "helping others" as her most effective method of coping with her experience, she uses pulse reading, needles, herbs, and nutritional counseling to help her patients. Most recently, Harriet wrote a book about Chinese medicine, another of her attempts to extend her compassion to the "millions of people, like mothers of children who never get better."

Ironically, the commitment to service that springs from technology survivors' hearts and hands occurs in the midst of a society that did not care enough about them in the first place to ensure their safety, and since the onset of technology-induced illness, has not cared enough about them to address their needs. In response, many technology survivors passionately build networks, from Gilberto

Quintana's individual acts for his neighbors in Pecos, New Mexico, and Nathan Robinson's for blind people in Vallejo, California, to the interlinking of asbestos, radiation, aspartame, DES, Dalkon Shield, Agent Orange, and pesticide survivors all over the country. Many survivors blossom as they develop their capacity for connection, and this blossoming enhances not only their lives, but the lives of everyone it touches.

HELPING THE
FUTURE

We have to get it stopped immediately.
We want no more victims of any kind.

—**Jesus Rives,** chapter coordinator,
Asbestos Victims of America

June 2, 1987: A New York businesswoman hikes onto Vandenberg
Air Force Base and beats the nuclear-weapons-related computer
NAVSTAR with a crowbar, bolt cutters, hammer, and cordless drill.
She is moved by the belief that it is her duty as a citizen to intervene
when her government's technology is "preparing a means of geno-
cide."[1]

June 3, 1987: Along a highway between Stinson Beach and
Tomales, California, protestors wearing gas masks block Department of Transportation crews spraying herbicides. State officials say
they'll keep coming back to spray. The citizens say they'll keep
returning until the herbicides undergo full testing for links to cancer,
birth defects, and other illnesses.[2]

November 14, 1987: A woman kicks the screen of an old television set at Sproul Plaza on the University of California, Berkeley,
campus. Other students wield mallets to smash a pile of TVs. This
is a student protest, 1980s-style, a self-described "act of therapy for
the victims of technology."[3]

August 21, 1988: Farm labor leader Cesar Chavez ends a thirty-
six-day fast by breaking bread with Robert Kennedy's children. Also
attending are Ethel Kennedy and Jesse Jackson. The fast is a protest
against the use of cancer-causing pesticides on table grapes.[4]

Such acts echo the British Luddite movement of the early nineteenth century when workers in the wool industry rose up in rebel-

179

lion and violently smashed the new shearing frames and gig mills. Most people think of Luddism in simplistic terms as vandalism by recalcitrant workers. In fact, though, it was a complex social phenomenon, an ideological struggle in which workers saw what owners were calling "the introduction of improvements" as a threat to long-lived social relations valuing the unity of family, community, and livelihood. They also favored the old, relatively grass-roots economy over the more hierarchical, expansionist industrial capitalism.[5]

Acts of technological protest by students at Berkeley and by people like Cesar Chavez also echo a more recent proclamation, the words of Free Speech Movement leader Mario Savio in 1964: "When the operation of the machine becomes so odious, makes you so sick at heart that you can't take part, you can't even passively take part, you've got to put your bodies upon the gears and upon the wheels, upon the levers, upon all the apparatus and you've got to make it stop."[6]

According to a California Department of Food and Agriculture study on how people assess "the operation of the machine," political participation among the general public does not necessarily reflect people's depth of concern. More people are anxious and fearful about dangerous technologies than are willing to join a community group, lobby an elected official, contribute money, or smash a military computer. The study asks: Who cares? Who acts? It answers: Many people care. Few act.[7]

Many technology survivors care, and many act. I was surprised to learn this fact. I expected that the demands of physical illness, in most cases, would prevent even the thought of involvement. Yet many survivors I interviewed indicate their most effective means for coping is political action.

The motivating factor is clear. Technology survivors *know*—in the most intimate and compelling way—what dangerous technologies can do to life. They know the disruption, loss, and uncertainty. They feel the breach of trust, and these experiences can catalyze them to question accepted beliefs about technological progress. Many generalize their personal knowledge of a particular technology to a broader awareness of the effects of modern technology as a whole. A DES son does not see diethylstilbestrol as an isolated technological failure. An environmentally ill woman does not see synthetic carpets as the only problem. They see them as symptoms of a whole system gone awry.

"What I learned is that our technology is killing us," says former asbestos worker Nathan Robinson. "I just hope people can start living on this planet earth without so many chemicals. The old people lived, and they didn't have so many chemicals. When I was a kid, I never heard of them. I've learned so many new words in the last twenty years. Not just asbestos, but dioxin, radiation, microwave, ozone. I know these now because I've been exposed to them. *Everybody's* been exposed to them."

Sarah Pirtle echoes Nathan's view. "A drug like DES is symbolic of what is happening throughout society today," she says. "With DES there is this horrible, negative heritage to pass along, and it's an emblem of all the violence to life—all the murder and maiming, all the toxic assaults on nature, all the rape, all the warring, all the fear and hatred—that is handed down, literally handed down to our children."

Gay Ducey's exposure to dioxin has taught her about the politics of the American social system. "We were fairly political before," she explains, "but we became infinitely radicalized by our experience. We saw our personal horror story as an outgrowth of a system based on aggression, acquisition, and alienation. In the midst of this, we confronted what Hannah Arendt calls 'the banality of evil.' Individually, the corporate people are nice enough. They love their families. They don't run down pedestrians. But collectively there are evils in our society. The corporation is an amoral force. It is a profit-oriented entity and has no interest in anything else, not you or me. This realization increased our desire to insist on justice."

Laura Martin-Buhler connects her experience of radioactive fallout not just to the system that creates it, but also to its impact on global ecology. "The worst part is knowing that children are dying of leukemia. The next thing that bothers me is knowing that dairy cattle are eating the grass and making the milk and feeding the children. This angers me! It defies my comprehension that there are humans who feel they can pollute the earth! Do they think that pollutants just go up into outer space and dissipate? What really happens is they stay and give us cancer. They destroy the ozone and give us cancer. Some people think the poisons wash out to sea and we don't have to worry about them. But in fact they destroy the plankton and algae, and then the big animals have nothing to eat and they starve. Just dump all the oil and sewage and radiation, and the

181

ocean will take care of it, right? But let's face it. The ocean has *had* it! The ocean is becoming a graveyard!"

With personal and often agonizing knowledge of the deathly effects of many technologies and a systemic view of the problem, many technology survivors reach out to help not just themselves or other people, but society as a whole—and, in their view, the future of life on earth. When I interviewed technology survivors, I learned that many women made sick by reproductive technologies fear that with the export of technologies like DES, IUDs, oral contraceptives, and birth control implants to women all over the world, the human capacity to reproduce without complication is threatened. Many people exposed to asbestos believe that enough asbestos is in circulation to give everyone alive lung disease. Many people exposed to toxic chemicals feel that the chemical waste already created is enough to destroy all life, and some radiation-exposed people indicate there is enough nuclear material in the atmosphere to kill us all. Jose Luis Roybal puts it succinctly: "Unless the mess is cleaned up, there isn't going to be anything left."

Against this backdrop, the effort to work for a better future can be psychologically healing. According to researchers, one of the most common reactions of those who undergo disaster is the fear that it will happen again. After events like hurricanes, tornadoes, and wartime combat are long over, people tend to harbor fear that they will resume.[8] In the case of a world overwhelmed with dangerous technologies, such fear is not merely a psychological exercise. The Dalkon Shield user indeed may be contaminated by electromagnetic radiation, pesticides, or benzene in a future situation. The Times Beach resident may flee to a new neighborhood only to find it has radioactive groundwater. Working to change the ideological and institutional system that perpetuates these dangers, then, offers the survivor a chance to envision a different, safer future.

It also provides one aspect of a healthy worldview that has been lost: a sense of continuity. Constantly faced with threats to the future in work interruptions, financial pressures, and the possibility of disabled reproductive capacity, defective children, and death, the technology survivor re-creates a sense of continuity by working for a better future. As Sarah Pirtle so passionately asserts, "You have to be part of breaking the spell. Because you've experienced the technology, because you've been a victim, you're in a unique position to lead

182

change. If you become cynical and numb, you're letting the technology have the last word!"

Most of all, acting to usher in a better future gives the survivor a renewed sense of heroism. It advances a sense of connection to forces and events greater than one's individual cares. It restores the all-important sense of purpose: In the midst of deterioration and death come acts of creativity and vision. With its call to participation and passion, helping the collective future transforms the survivor into a hero, a hero of the technological age.

Some technology survivors look to the future with acts that respond to their immediate contamination. Having endured years of illness and near death catalyzed by reproductive technologies, Wendy Grace is now looking ahead, healing and strengthening her body so she can bring forth a new world in the most literal way. She wants to have a child.

Others address the need to rebirth a new world in direct response to the particular technology that harmed their health. They fight to rid the world of its dangers. This approach is akin to the ecological adage that if all the cars in the United States stopped, the nation's roadways would, within months, be vanquished by flowers, weeds, and shrubs, reborn as natural landscape.

Maria Carriaga has worked to restore the natural landscape of the "sawdust neighborhood" in Albuquerque by becoming a community activist. The transition has not been easy. Having lived her life as a housewife and mother, initially she felt intimidated, but Maria's passion to save her neighborhood compelled her beyond the fear. She became a public speaker, negotiated with officials, and joined a movement that, so far, has enforced health codes, conducted public health studies, registered voters, and lobbied city officials, representatives, and senators.

"I have learned that people cannot sit back," she says of the struggle. "You have to go out there and fight. Take me. I'm going to be out there fighting till the day I die. Inform the people! A lot of people still don't know what is going on at Ponderosa. I lived a hundred feet from there, and *I* didn't know. You have to educate people about these chemicals. They don't know what they're breathing and drinking. Now I go out and tell them, 'Look at *me*. I don't want you people to go through this. I want you to learn, to get involved. Don't be afraid of the city people or the state people or the big industry. . . . And if you don't know what to do, find someone who does! There are organizations out there that can help. If you are puzzled and afraid, ask for help. Or you're going to wind up dead.' "

Maria's link to the future is perhaps most pointed when she speaks about the children. "When the company offered to buy my house, I said no—not just because it wasn't enough money to even buy a garage. We have to fight them. We have to fight industry. I mean, I might move away. But my grandchildren are still going to be here."

Working with the Southwest Organizing Project, Maria learned a crucial lesson in making effective social change. When she went on a tour of the particle board plant next door, she found allies *in* the factory. "We couldn't believe it," she explains. "The dust was falling like snow everywhere, and the men were sitting around eating sandwiches covered with dust and formaldehyde. And you know, that's going to affect their health too. I talked to some of the workers, and I said, 'Look, I'm sorry about your jobs because we have to protest this place.' The guys said, 'Madam, we look over at where you live every day. We *know* what you're going through. *Go for it!*' "

In Texas, Kari Pratt and Robert McIntyre are also engaged in a struggle to create a better future. After the vineyards began spraying pesticides that drifted throughout the area, Kari and Robert made a bold decision. Not only did they make an effort to help their neighbors deal with the illnesses brought on by spraying; they also decided to work to stop the contamination. They spoke with their neighbors in Fort Davis, informing them of the possible link between pesticides and encroaching illness, and they gathered medical research. Calling the vineyard owners as well as local and state officials, they tried to convince the offending businesses to switch to nontoxic, organic methods of pest control. Both Kari and Robert considered searching for a new place to live, but realizing that pesticide spraying is prevalent throughout the United States, they pledged themselves to stand their ground—for the future.

In a 1988 letter to the *Jefferson Davis County News,* Kari writes: "Is it a coincidence that after each known application of pesticides at Blue Mountain Vineyard reported by the Texas Department of Agriculture, numerous individuals in the Davis Mountains Resort and city of Fort Davis became ill?" She proceeds to report the medical effects of the chemicals used: paraquat, Bayleton, captan,

Benlate, Guthion, and Sevin; describes the sicknesses in Fort Davis; and concludes:

> I have lived in Fort Davis for four years. I came to this area to recover from lindane poisoning. I was nearly an invalid [when I came and had] almost completely recovered at the time the vineyard hired a licensed pesticide applicator last August. I became acutely ill again last summer after . . . the spraying of paraquat. . . .
>
> Texas Tech University informs me that grapes are the most highly and intensively sprayed agricultural crop in the United States. . . . When pesticides are sprayed, aerosoled, or dusted, there is almost always measurable drift. A U.S. EPA report notes that more than 99% of the applied chemical becomes unwanted residue on the treated crop and in the environment inside and outside the target area. The conclusion is inevitable: *Broadcast pesticide applications are an uncontrollable technology and must be reduced in favor of nonchemical resource management.*
>
> I still believe we in Fort Davis have the constitutional right to be free of poisons when not applied on our own properties. I do not believe there will be justice until there are *absolutely no illnesses created from pesticide drift.*[9]

Their struggle continues. They have met with Texas Commissioner of Agriculture Jim Hightower to inform him of the problem and work toward legislation to establish toxic-free areas and better testing of toxic substances. They have met with the Highway Department to try to convince them to stop spraying pesticides on the roads, and with some thirty other Fort Davis residents, they have initiated a lawsuit against the vineyards.

Like Kari and Robert, former Marine Victor Tolley is motivated by his desire for future justice. Now in his seventies and suffering from health problems catalyzed by exposure to radiation in Nagasaki, Victor embarked on a hunger strike in 1986 to protest the Warner Amendment, which forbids citizens from suing government contractors involved in weapons manufacture. After two weeks, his health deteriorated and his friends convinced him to stop. He continues to work for radiation survivors through NARS.

June Casey is also motivated by her desire for a better future. "Speaking out was a painful thing for me," she says. "I had lived forty years with the exposure, why did I have to talk now? But then I felt encouraged when I heard Dr. David Bradley at the NARS conference in 1986. He was the medical doctor at the Able-Baker nuclear tests on Bikini Island. He said, 'Sixty percent of the people exposed are no longer with us. You are the forty percent who has to speak for those who can no longer speak for themselves. *Never miss an opportunity to tell the story.*' So now I am trying to reach as many people as I can, and I'm doing it for the people who died prematurely. I'm doing it for myself because if I didn't, I would be in evil complicity with the morally depraved actions of the government. And I'm doing it, most of all, for all the people in the future."

Believing that any person is capable of "a moral awakening," June goes on TV talk shows to explain the health effects of radiation exposure. She sends photocopied articles about the nuclear issue to friends all over the country, and she has served as acting secretary of NARS.

Many organizations like NARS join individuals working for an awakened future by addressing specific technological hazards.[10] On a national level, the Citizen's Clearinghouse for Hazardous Wastes (CCHW) helps hundreds of grass-roots organizations to clean up toxic landfills, stop ocean dumping and incineration, and force corporations to discontinue producing more chemicals. In 1986, CCHW held its Fifth Anniversary of the Grass-roots Movement Against Hazardous Wastes in Washington, D.C., attracting some four hundred community leaders and concluding with objectives for the movement over the next five-year period. Also working nationally, the Campaign Against Toxic Hazards is waging a state-by-state organizing campaign to wield political muscle through the electoral and legislative processes. Its declaration of citizens' rights asserts the right to be safe from harmful exposure, to know about dangerous contaminants, to be compensated after exposure, and to enjoy the benefits of prevention, protection, and enforcement. Asbestos Victims of America works not just to gain compensation for its members, but also to stop the mining and manufacture of asbestos products nationally. Likewise, NARS, known for its work to gain medical compensation for radiation-exposed persons, is also against the production, testing, and use of nuclear weapons. Americans for Safe Food is organizing a national coalition to lobby for pesticide- and drug-free food and

186

establish national standards for organic farming.

Statewide coalitions to address toxic exposure issues include the Arkansas Chemical Cleanup Alliance, the Louisiana Environmental Action Network, the Maine People's Alliance, and Vermonters Organized for Cleanup. On a local level, the West County Toxics Coalition is attempting to press Chevron Corporation to adhere to safer environmental standards in its refineries near Richmond, California, an area where the cancer rate is extremely pronounced.[11] Albuquerque's Southwest Organizing Project is working both to stop chemical exposure to Chicanos living in industrial areas and to develop the community economically before more toxifying corporations buy up the land. The examples are countless, their fundamental message encapsulated in the exhortation: "Not in my backyard" ... or workplace ... or home ... or medical treatment ... or food ... or air. In fact, the movement as expressed by all these individuals and organizations fighting for the future unites with a single message: Not on our planet.

Against the backdrop of a flurry of effort for a future unburdened by specific technological hazards, technology survivors are also arising to raise public consciousness about the system creating the hazards. Many see that basic underlying thought patterns and institutional structures must change if the problem in its myriad manifestations is to be solved. Many point out that human survival needs can be met in less damaging ways and that human psychological needs expressed through and seemingly answered by technological development can be addressed, as they have been in other cultures, by more life-enhancing practices. Technology survivors are also aware that the problem of modern technology encompasses not merely a case of leukemia here or an immunologic disorder there. At this moment it encompasses a threat to the survival of all life on earth.

THE
FUTURE

And now we sing
together, violin
and pipe,
we sing what we
all know, tambourine,
tambourine, tambourine
and pipe, *what we*
all know, tambourine
tambourine, tambourine
and pipe, *what we*
do will bring
us danger,
tambourine,
we are taking
great risks, violin
and pipe, violin
and pipe.

—**Susan Griffin,** *Woman and Nature:*
The Roaring Inside Her

THE EARTH IS A SURVIVOR TOO

The animals are getting immunological diseases too, the seals, the koala bears, the whales. You can't always see it, but the trees are getting weaker from the roots up.

—**Rhiane Levy,** sensitized from overexposure to electromagnetic radiation in computers, fluorescent lights, TV and radio waves, and other electronics

In December 1972, NASA's *Apollo 7* spacecraft was headed for the moon, and on that voyage a small event occurred that proved exceptional in helping modern peoples realize we live on a single, shared—and limited—planet. The first clear photograph of the whole earth was taken.[1] Since then, this image has appeared and reappeared around the world in newspapers, books, and films and on television. It is a picture of our home, in its full, vibrant splendor.

Ancient and native peoples have long believed that this planet is a living being, female in gender. They have called the earth Great Goddess, Earth Mother, and Gaia. But looking now beyond the beauty of her swirling cloud patterns, we detect something else. The earth is bearing the hard, gray armor made of the steel and concrete of modern technologies. Once green with possibility, she now whirls through space in a vapor of petrochemicals and plutonium, emanating an unearthly aura of microwave, radar, and electromagnetic pulses. The earth is a technology survivor too. Look more closely.

The surface of the earth is dotted with cities. Everywhere you can see them. Enveloping the countryside with their freeways, parking lots, and skyscrapers, spouting sulfur dioxide from smokestacks and carbon monoxide from vehicles, cities are the busiest machines on the planet. Bombay, Paris, Los Angeles, Rome, Singapore—they

191

superimpose themselves upon the earth like mock attempts to re-create her own sacred power spots. Their extended corridors of concrete, wire, and microwave reach out to other cities until, altogether, they threaten to wrap the planet's body in one vast urban machine.

Every technology the *Apollo 7* glimpses from its celestial vantage point flows from or relates back to the cities. You can see the pink lights and cooling towers of the nuclear power plants that provide energy for their operation. Generating one twelfth of the world's electrical power,[2] reactors arise out of the land like inviolable shrines. There are 350 of them[3]—in Brokdorf, West Germany; Bilbao, Spain; Hanford, Washington; and Tsuruga, Japan.

Also providing the energy required to run the cities are the world's electrical power grids. From the *Apollo 7* you can glimpse high-voltage transmission lines running across the land nearly everywhere people live. In the United States alone you see over 365,000 miles of overhead lines.[4] You see thousands of power stations generating the electricity to fuel these lines—twelve hundred hydroelectric dams, twenty-two hundred oil- and gas-fired plants, and nine hundred thermal plants.[5]

Scanning from this photographic post in the sky, you notice vast territories blighted by the brown and gray of pollutants, you see the factories on every continent burning fossil fuels, smelting metallic ores, and discharging industrial wastes. Each year human activity injects 100 million tons of sulfur dioxide, 35 million tons of nitrogen oxide and 5 billion tons of carbon monoxide,[6] into the atmosphere.[7] The result: Whole portions of the planet's surface are wasting away. Do you see? In East Germany they call it *Waldsterban:* "forest death." The trees are no longer hearty green but black, withering away from the acid in the air and rain. The plants are sick and dying, and everywhere their loss makes the soil less fertile. To ensure food growth, people are wielding axes, chain saws, and bulldozers to clear more trees and farm more land. They are applying more chemical fertilizers and more pesticides—causing more acidity and more dying forests.

Around the planet—in Africa and Guatemala, in North Carolina and New Zealand—chemical fertilizers are used: in 1984 alone, 121 million tons.[8] Pesticides are sprayed on farms and forests: 7 billion pounds each year.[9] Zero in on southwestern Texas. In the Davis Mountains you can see Robert McIntyre standing by his flatbed truck. He is tired and dizzy from the paraquat in the wind.

Plastics are everywhere. You can see them on the side of the road. They are leaching into the groundwater from landfills and releasing hydrogen chloride during incineration. The merchant marine dumps 639,000 plastic containers into the sea in one day,[10] and production of synthetic chemicals reached 320 billion pounds in 1978.[11] In fact, new chemical compositions whose effects on the body of the earth are unknown are continually released. Fifty-five thousand human-made chemicals are now in commercial production, and each year close to a thousand more are introduced.[12]

With all these pollutants circulating on the planet, it is no wonder her blood fouls. Look to the Danube River. It is no longer blue, but brown. The Cuyahoga River in Ohio was declared "dead" in the 1970s, the Ganges is polluted, and each year Niagara Falls drops 100 tons of toxic chemicals over its rocks.[13] Look close up again. There is Sarah Pirtle in a blue bathing suit walking along Marblehead Beach in Massachusetts. She is showing three-year-old Ryan the sea shells. They come upon a heap of dead fish, a broken catheter bag, some sandy vials of blood, a pile of hypodermics, their needles bent and broken. Look closer. In an aluminum boat in the middle of Torch Lake, Michigan, Jane Woolf reels in her first catch of the day. She removes the fish from her hook. It has cancerous lesions around the gills.

Scan the whole planet. There's the African continent. Whole portions are turning barren and dry. Since Western colonization—and technology—have disrupted tribal life, native populations have swelled out of proportion. People are chopping down more trees for firewood and farmland. In eleven out of thirteen West African countries, the demand for firewood has outstripped sustainable yields.[14] With the trees gone, the soil deteriorates. It washes away with the rain and blows away in the wind. There is less evaporation from the land, and rainfall dwindles. Water tables fall. Wells dry up. Look to Ethiopia and Mauritania. There is almost nothing left of life.[15]

As you scan the globe, you notice dark veins of concrete, tar, and steel crisscrossing the earth's belly. These are the world's thoroughfares. They transport the construction materials and petrochemicals, the nuclear weapons and irradiated fuel, the commuters and vacationers of technological society. There are the railways: 811,622 miles of route intersecting the land with their tracks and coal-firing engines.[16] There are the automotive highways: in the United States alone, over 4 million miles of road[17] accommodating 348.5 million vehicles.[18] In the Soviet city of Togliatti, huge patches of greenery are wasting away from carbon monoxide. Mexico City's trees wither along the most traveled roadways.[19] Look to the South American continent. In Brazil the Trans-Amazon Highway plunges hundreds of miles into what a decade ago was vital rain forest. With the onslaught of civilization, great numbers of botanical species are disappearing—according to 1985 NASA photographs, 11 percent so far.[20] Note the yellow bulldozers mowing the rain forest for development of cattle ranches. If current trends persist, a desert is predicted there by the year 2050.[21]

Looking at the planet, you can also see labyrinths of pipelines bearing oil and gas. They travel offshore in the Gulf of Mexico and run across the shifting deserts of the Middle East. They run from the North Sea to the British mainland, from Algeria to Italy, from Siberia to Western Europe. In the United States you can trace 227,000 miles of pipeline.[22]

Scan now for the nuclear weapons industry. Its design laboratories and production facilities speckle the earth like plates of armor. Across the United States you can see an elaborate network of them in Rocky Flats, Colorado; Savannah River, South Carolina; Fernald, Ohio; St. Petersburg, Florida; Amarillo, Texas; and Kansas City, Missouri.

You can't see the weapons, though, despite the fact that there are some fifty thousand of them.[23] They are hidden in underground silos, loaded on bombers, and circulating on trains. They are in the plains of North Dakota, in central China, in the Sea of Okhotsk, under the icepack of the Arctic Ocean, on the Mediterranean Sea.

Looking more closely, you glimpse the earth turning. There is Pecos, New Mexico. You can just make out Gilberto Quintana limping out to the old wringer washing machine in the backyard. His body begins to tremble. He falls down in a convulsion. A thousand miles west in San Jose, California, Andy Hawkinson is being wheeled on a trolley into the emergency room of the hospital, tubes entering his mouth, nose, and veins.

There is more to see. What are those antennae protruding from the countryside? They are the military command centers—vast computers coordinating the actions of the world's arsenals. They lie below the surface of the earth, like the underground center at Cheyenne Mountain in Colorado, "hardened" with tons of concrete and connected to civilian communications centers by equally "hard" cables and microwave relays.

As the earth rotates on its axis, you notice odd structures. In the middle of the Pacific Ocean, you can see an enormous concrete disk flattening an uncommonly treeless island. This is a nuclear dump, not to be disturbed for over 100,000 years. Look to Washington State. Here billions of gallons of radioactive wastes are hidden below the ground in steel tanks, unlined trenches, and ordinary ponds.[24] Such tombs can be detected in New Mexico, Siberia, France, and India—wherever nuclear facilities operate.

There are other waste dumps festering like carbuncles beneath the earth's skin. During the 1970s, 35 million tons of toxic wastes were generated each year in the United States,[25] while 1980 saw a leap to 63 million tons—enough to fill three thousand Love Canals.[26] The Environmental Protection Agency tells us that as many as fifty thousand waste dumps exist in the United States.[27]

You notice another strange phenomenon. Far from the cities, you notice the earth rumbling and shaking. Then a huge plume of dust arises with no visible cause. These are underground nuclear explosions. Look to Siberia, the western United States, the South Pacific, the Sahara Desert. Like lethal assaults against her life, they gut the earth's insides and vent poisons into her atmosphere, soil, and groundwater. There have been 1,250 such eruptions since 1945,[28] over 250 directly into the air and nearly 1,000 underground.[29] Through the dust in the wind you can see Laura Martin-Buhler resting at a picnic table in Kanab, Utah. She lifts a bottle of pills from her purse and methodically counts out two.

There is more. Mining technologies penetrate the body of the planet. To provide the staples of technological society, they extract heavy metals, radioactive minerals, coal, oil, and asbestos as if excavating the very bowels of her body. In Bolivia, South Africa, and Zambia, you see drills invading her interior, bulldozers and dragline cranes scraping and pummeling her skin. You see conveyor machines and processing plants readying the materials for human use. You see the coal slag heaps, uranium tailings, and stripped land left behind when the miners depart.

Look now to the seas, majestic and blue. Covering 70 percent of the earth's surface, they too are barraged by technology. Supertankers carrying 3 million barrels of oil ply the oceans,[30] some of them cryogenic ships transporting liquid natural gas. There are offshore platforms and rigs along the coastlines. There are drill ships, surveyor ships, tugboats, submersible vehicles, and support ships to carry fuel and pipes to the platforms. You see freighters transporting consumer goods, resources, and irradiated fuel. You see yachts and ocean liners carrying vacationers across the seas.

Fishing is no longer an activity for boys toting bamboo poles or men with hemp nets. It is now conducted with gigantic fish-processing factories floating beside fleets equipped with sonar tracking stations, radar, computers, and high-tech ship-to-shore communications. All of these increase the efficiency of fishing so that from 1950 to 1970, the worldwide catch tripled. It now exceeds the fishes' capacity to reproduce.[31]

Warfare hardware pervades the seas. At any moment, thirty-five missile-carrying submarines are diving and surfacing in the oceans. Together, they carry a total of thirty-one hundred nuclear warheads.[32] See the iron warships cruising the Persian Gulf and the aircraft carriers in the Sea of Japan. Note the hydrophones bobbing with the waves, collecting acoustic information for submarines. See the lasar and infrared detectors and the surveillance processing facilities in the waters all over the earth.

At the naval shipyard in Suisun, California, a cluster of sad-faced people is boarding the U.S.S. *Clamp*, a ship that was irradiated during the Crossroads nuclear tests near Bikini Island. The people carry a banner: "National Association of Radiation Survivors." They sing and pray. After forty years, the ship is still radioactive.

There are also technology-produced poisons in the seas. You can see them. Relatively clean waters flow into the Mediterranean at Gibraltar and quickly become darkened by pollutants from the Adriatic and Aegean seas, from the Rhone and Nile rivers. The Red Sea and Hiroshima Bay are dying. Each year, tanker spills, automobile emissions, and polluted rivers dump 10 million tons of oil into the oceans.[33] The United Kingdom's reprocessing plant at Sellafield pumps 1.2 million gallons of radioactive wastes into the sea each day.[34]

The earth's atmosphere is cluttered with metal objects. Notice them floating and darting above mountain peak and valley. They are helicopters, satellites, missiles, airplanes. One is a sleek jet taking off from Denver. Its destination is Minnesota, home of the Mayo Clinic, and one of its passengers is a pale-looking Sukey Fox. Look more closely. Chemical antibiotics flow through her veins. There is a piece of plastic still lodged in her abdomen.

See the rockets arching into the sky. Five hundred are launched each year. Then there are the space stations and spacecraft—all backed up on earth by launch sites, communications terminals, tracking installations, and data processing centers. Between 1958 and 1976, the United States, the Soviet Union, France, Britain, China, and NATO launched a total of 2,311 military and nonmilitary satellites into the atmosphere.[35] Some, like reconnaissance satellites, fly 90 to 300 miles above the earth. Others, like communications satellites, circle at up to 22,000 miles, each able to transmit thirty thousand telephone calls at once.[36] Together with all the lost wrenches, frozen human wastes, spent payloads, and rocket bodies, there are some fifteen thousand human-made objects whirling in planetary orbit.[37]

But the earth's atmosphere is not only cluttered by metal objects. It is also dirty. You can see the pink-and-brown clouds from coal-gasification plants and automotive exhaust encircling the planet. What you can't detect are the microwave and radar rays flooding the atmosphere from communications centers, radio towers, and satellites. You can't see the electromagnetic fields emanating from the thousands of miles of power lines. Nor can you see the 150 million tons of radioactive materials released into the earth's stratosphere by nuclear manufacture and testing.[38] But they're there. The very photograph of the earth that inspires this exploration, ironically, was taken by remote sensor technologies beaming microwaves at the planet.

Our planet is girded at every curve by human technologies. But there is more. If you sit long enough in your photographic mooring in the sky, you can see unearthly happenings. A vast explosion rocks the Ural Mountains and leaves thousands of acres radioactive. Coal smog envelops London, England, and Donora, Pennsylvania. Mexico City is lost from sight. Motorists in Molbis, East Germany, are using their car headlights at noon to see through the pollution. A bomber drops a nuclear weapon on the Spanish countryside. A satellite crashes in northern Canada. A tank of jet fuel explodes in Japan. Five thousand gallons of radioactive wastes leak into the earth at Hanford, Washington. A rocket explodes over Cape Canaveral. A cloud of methyl isocyanate escapes from a plant in India. A reactor in the Soviet Union spews radiation around the globe. A newborn child in Kraków, Poland, dies from air pollution.

The earth is in pain. She grieves her losses and wounds. Her future is uncertain. The earth is a technology survivor too.

WHAT TECHNOLOGY SURVIVORS WANT TO TELL US

> Look at the technology. Look at the quality of life that comes from it. Look and see if that technology is really worth creating.
>
> —**Rhiane Levy,** sensitized from overexposure, to electromagnetic radiation in computers, fluorescent lights, TV and radio waves, and other electronics

In light of the tenuous condition of our earth, technology survivors have a special mission. They present the most striking personal illustration of the consequences of modern technological development. And they have the most passionate voices to describe these consequences. According to pesticide-poisoned anthropologist Joan Wescott:

> For those who have been disabled by chemicals and who are now sensitized to pollutants to an incapacitating degree, it can become part of our survival to accept our affiliation as a ministry. . . . We are the bearers of the bad news, and so we are the outcasts. But throughout history it has been the outcasts—the victims—who have the great potential for prophesy and vision.[1]

Their prophesy begins as a challenge to the psychological defense mechanisms that keep us from thinking about technological threats. "Watch out!" asserts Dalkon Shield survivor Sukey Fox. "Be informed," warns Charlotte Mock, whose water was contaminated in

Albuquerque. "Don't believe everything your doctor tells you," explains Diane Carter, also a Dalkon Shield user. "Question everything!"

On the surface, the key to surrendering defense mechanisms about technology-induced danger is simple. It is the admission of vulnerability. More than other people, technology survivors feel vulnerable before the forces of technological development and use. The repression they formerly maintained transmutes, through their experience, into a new awareness: "I am painfully aware of the hazards." Denial turns into "I can no longer pretend it isn't real." The projection they used to employ translates into a new statement: "No technology can give my life meaning or save me." For the technology survivor, introjection becomes "I trust my own perception." Personal disconnection becomes "It *is* happening to me," rationalization "I can face the facts," and selective inattention "I see the crisis in its totality."

Such psychic breakthroughs, while undergone by technology survivors out of necessity, are required for us all if we are to muster our efforts for collective survival. This is the most impassioned message that technology survivors want to tell us. With their hearts and minds, they understand the implications of *not* making such changes in consciousness about technology, and they understand the personal challenges involved in making them. Above all, they know that such changes demand courage and creativity. After making them, technology survivors have been left with a sense of fragility in technological society. They have been left exposed before the uncertainty of the future. They have also been left in a state of unqualified honesty. Such a state can become a source of strength and transformation. Stripped down, as Joan Wescott says, to "the ground rules for essentials,"[2] technology survivors are open to new perceptions and new ways.

One of the questions I asked the forty-six survivors I interviewed was: "If you could tell the world one thing about your experience, what would it be?" Their responses are testimony to the learning they have gained. Technology survivors want us to know that the prevailing beliefs about technological development and use serve neither life nor hope. The notion of technological progress is the first to go. "Technology hubris!" cries DES mother Pat Cody. "What a price we pay for all the so-called scientific advances!" "Being a technology victim infuses you with trepidation," reports downwinder Laura

200

Martin-Buhler. "You don't embrace every new 'advance' and call it a Great Achievement." Adds birth control pill user Meryl Tavich, "I feel older and wiser, and as a society I wish we felt older and wiser. We can no longer rely on scientific 'progress' as the ultimate bringer of truth."

Nor, in many technology survivors' minds, is the primacy of the technological solution adequate. "We Americans think there's a technological solution to everything," says Teddy Ostrow, who was exposed to toxic chemicals as a hydrologist. Adds DES daughter Sarah Pirtle, "The spell about technology, it's everywhere! It's in the glossy ads about medical breakthroughs. It's on television when helicopters and computers save the day. The chemist is so in love with his invention, he doesn't want to evaluate if it's harmful. But look what's happening! We're dying! We have to break the spell."

To many technology survivors, the ethic of human mastery over nature is revealed as ultimately destructive. Because the "spell" is already broken for them, the daily technological disasters and accidents, from oil spills in Antarctica to radioactive transport spills in France, become painful reminders of the folly of the myth of control—and the tenuous state of life on a planet whose fate is increasingly controlled by technological development.

"We shouldn't be tampering with Mother Nature,' " says atomic veteran Andy Hawkinson. "But that's just what's being done. And when the trade-off is pain, anguish, and life—it's *awesome.*" "And absurd!" adds downwinder June Casey. "Scientists and politicians are determining whether we live or die." "Western civilization has thought it has carte blanche on nature," says Susan Griffin. "Its technologies have been developed with such a lack of understanding of the complexity of nature. You can't intrude on a natural system without consequences."

Many survivors, by generalizing their experience of dehumanization through a technological event, come to see that the accepted social belief in the mechanization of life is responsible for mass dehumanization. Says formaldehyde-poisoned Maria Carriaga: "There is no caring anymore. I come from a big family, and it was so different then. It's sad to say, but it's going to be a long time before we can live together again."

Through her experience with iatrogenic disease, Susan Griffin has come to understand the extreme mechanization of health care. "The education of doctors has to change," she says. "They need to understand the body is not a machine with interchangeable parts. They need to realize that studies in the *AMA Journal* just reveal the *known* version of the truth of the moment. But instead the doctors keep trying to fit their patients into this grid or that grid. Real scientists understand that these studies are just little windows poking through to a reality that is far vaster than any experiment can show. I think doctors' education should emphasize that science is just a tool. Healing is much larger than science. Real healing involves changes in an organism on all levels, on the psychological and spiritual levels which cannot necessarily be measured or pigeonholed, as well as on the level of blood chemistry and brain waves."

Sarah Pirtle says that mechanization in the scientific laboratory makes it "a cold, metallic existence." "Workers are more involved with flow charts, research deadlines, and abstract data," she says, "than they are with human consequences." Adds Charlotte Mock, "My nephew is a scientist at Sandia Labs where they design nuclear weapons. He has *always* worked there ever since he finished school. What kind of people are these? Many of them cannot exist outside that laboratory because they've no experience with life."

Electronics worker Irene Baca finds the same standardized "assembly line" approach in industry. "Companies have got to start thinking of their workers as human beings, not just moneymaking cogs in a machine. They've got to look at us like we are their brother or mother, somebody they care about in this world."

When I asked interviewees if they have a vision of a society that uses technology in safe, beneficial ways, I discovered that dissatisfaction with current technological ideologies leads many to contemplate a different kind of world, a world in stark contrast to the one they know. Not surprisingly, their visions derive directly from their own positive experiences of coping. The survivors I interviewed paint a picture of a society based on a sense of personal responsibility, respect for spiritual reality, service to others, and concern for the future. All of these qualities derive from the remembrance and reinstitution of a simple and seemingly lost human quality: *caring*. Together they present a new and decidedly nontechnological foundation for modern society.

SELF-HELP. One way technology survivors have invented to cope with technology-induced disease is by taking bold steps to help themselves. In the face of unwanted life events, they have taken responsibility for their emotional lives, imbued their experience with meaning, informed themselves, made crucial health care choices, and created support communities. Several interviewees want to tell us that for positive change to take place in society, a similar sense of personal responsibility must prevail—not only among those obviously harmed by technologies, but also among obvious and not-so-obvious perpetrators of technological disaster. June Casey feels that the first step toward building a better world must be the admission of past error by those responsible for the errors. Speaking specifically about the nuclear weapons situation, she says, "If the government is sincere about changing, they must show their sincerity by revealing the records about what has been done to people, where the accidents have taken place, who's been affected, and why. Of course, like Watergate, there will be the inevitable illegibles, deletions, and white-outs—so we'll never really know everything. But if the government is sincere, the first step is to release all documentation and admit responsibility."

Thérèse Khalsa, who sustains brain damage from a commercial product, extends the idea to corporate responsibility. "It's not until corporations publicize the harmful effects of their products that people will even have enough knowledge to make headway." Chlordane-poisoned Bliss Bruen suggests "a national course in mistakes: how government agencies, corporations, and scientific institutions have perpetrated dangerous technologies even when they've been totally in the dark and have acted as if they're the know-it-all watchdogs."

From their own experience, technology survivors are aware that if such admissions were to take place, shock, anger, sorrow, fear, and relief would follow in a collective grieving stage, a mass expiation in which citizens would let go of old beliefs in preparation for what Anika Jans calls "the crucial turning point." This is the moment many technology survivors dream of: the time when people in mass, technological society would consciously and willfully embrace responsible action for survival. Democracy is a recurrent theme. Many technology survivors feel that their predicament was caused by a lack of democracy in technological development and use. As a result, their view of democracy reflects a basic principle of self-help: Each citizen and each institution actively shares awareness and responsibility for social well-being.

Such a vision of mutual responsibility would involve the forging of new, less hierarchical relationships within society, relationships based on participation by all parties. According to Meryl Tavich, "Both doctor and patient are responsible. The doctor has some tools to offer, but the patient is also a participant. You need to be informed, and the doctor needs to listen to you." "Everyone on the job should watch out for safety," says Irene Baca. "*Everyone* should make it a safe, clean place."

A truly democratic vision would also include the relationship between citizens and government. Both Bliss Bruen and Teresa Juarez speak about the need for a grass-roots democracy in which people know about and have a say in what happens to them not only in accepted areas of democratic participation like voting for representatives, but in areas like technology production, governance of the workplace, and use of economic resources.

"People are not informed!" explains Teresa Juarez of the Southwest Organizing Project. "We don't even know what we're up against, and this is wrong. If a company is coming into your community, you should have the right to say something about it. There has to be a mechanism so we have more control over dangerous technologies. And I'm not only talking about Chicano people. I'm talking about all races because toxic contamination crosses all lines. It has no boundaries. The most important thing at this point in history is the people's caring and involvement. They care about their lives. They care about their children. They care about their homes. They care about what's happening to them—and they care enough to become organized to make change."

"My vision," echoes Bliss, "is that we'd have such an active and informed citizenry that we'd be powerful enough to influence change. As a society, we'd be putting our resources into solving problems instead of creating them. What lies between where we are now and this vision is a major educational and empowerment campaign."

GOD'S HELP. Technology survivors also cope by opening themselves up to an expansion in awareness from identification with ego and physical needs to identification with the life energy itself. Some express the idea that the drive to technological excess stems from such a lack of spiritual attunement—and the kinds of technologies we have today, by attempting to control and remove us from nature, only serve to further the limited ego identification that then fuels our psychological need for more technological development. "Our machines reflect our society's worldview," explains Sarah Pirtle. "It's a worldview of mechanization, alienation, and violence against life that creates more mechanization, alienation, and violence—not love for life."

Many technology survivors want to tell us that expanding our identification beyond ego and physical demands "not only gives us guidance for grappling with personal and social change, but also lays the basis for a better, more responsive society," according to Anika Jans. "The most important action at this point is prayer," says atomic veteran Jose Luis Roybal. "We have to unify before we can work on our problems and help one another."

Adds Sarah Pirtle, "If we can realize that we are so much greater than our technological ideology suggests, we can grow like a garden instead of always forcing ourselves to control life." "If we understood how precious life is," reveals Wendy Grace, "our society would reflect caring for life." In other words, technological development and use would reflect a different set of values—for June Casey, one in which "every human is worth protecting"; for Thérèse Khalsa, one in which we "surrender the extreme drive to control life and learn to trust"; for Susan Griffin, one in which we "live in the natural world."

According to interviewees, one important shift catalyzed by such life-affirming values would take society away from its uncompromised insistence on scientific-derived truth to include the validity of subjective truth. "Our very survival is up for question. We have to trust our intuition in unknown situations," explains atomic veteran Ricardo Candelaria. "How else can we get ourselves out of this dead end?" "In this culture, we're trained to give over our self-knowing to the experts," says Harriet Beinfield. "One of the lessons we've learned is that it doesn't work, it never worked, and we each possess the power to know more than we think we know. A shift back to acknowledging this power could only help our society to detechnologize and rehumanize life."

Another social shift many technology survivors advocate would takes us from a fragmented, mechanistic experience of life to one of interconnection and wholeness. "We're all related to each other and to the ecosystem," Harriet explains. "My life depends on your life. My life depends on the rain forests in the Amazon. My life depends on the microorganisms in the ocean. Our society has to make changes at its core to reflect this interconnectedness. Now everything is about separation and alienation. We send unwanted pesticides to Third World countries and then, surprise! The poisons come back to us in coffee, bananas, and cotton. What if the goal was to organize society around what's good for life?"

Last, an important shift would take us toward understanding the process of spiritual growth as a crucial unfolding not only in an individual's life, but in human evolution. "Despite how bad it is out there, the opportunity for love and contact brings hope," says Wendy Grace. "I don't know anything for a fact, but our human capacity for growth gives me deep hope for our collective ability to change."

HELPING OTHERS. Technology survivors cope by forging a sense of heroism and community in service to other people. Many of those I interviewed expressed their understanding that technological society depends on institutions that do *not* value human caring and that the harm done to them originated in this kind of alienation. They feel that service to others is a forgotten quality that—if rediscovered, valued, and institutionalized—could become a crucial aspect of building a better society. The desire for service as a socially sanctioned activity is, in part, a plea for a more humane, cooperative society, one that exudes a sense of safety and caring. "What life comes down to in the end," says pesticide-poisoned Robert McIntyre, "is how one person treats another person." "It's hard to hurt someone you see and talk to and care about," asserts Asbestos Victims of America's Heather Maurer.

Sarah Pirtle has a number of ideas for helping other people. One is to help the creators and disseminators of technologies—the scientists, engineers, politicians, and entrepreneurs—by keeping them in daily contact with people from all walks of life. In her vision, they could meet in "affinity circles" or "teams" to keep in touch with what she perceives as the purpose of today's society: "healing." Sarah suggests rewards rather than ostracism for "the brave whistleblowers who help us identify the dangers," and as a mother and educator, she proposes educational programs to enhance young people's understanding of "empathetic fields of knowledge, cooperation, and respect," providing a humanistic balance in a society based on scientific knowledge, competition, and disregard for human needs.

Technology survivors also want us to know that they need help. "We need relief from pain," says atomic veteran Gilberto Quintana. "Society should come to the aid of the victims," says DES daughter

Perry Styles. "I still go to the Veterans Administration every week," explains Ricardo Candelaria, "and every week they tell me to come back another time. It is only right we not be forgotten." Wendy Grace's vision of a better world begins with sick people meeting in "healing circles" to help and care for one another.

From their own experience of isolation and then reconnection through service to other people, many technology survivors feel that service could be the glue for connecting human relationships in a new kind of society, one based on care and respect. The impulse behind its enactment is simple. Says June Casey: "Affirm life."

HELPING THE FUTURE. Last, technology survivors have learned to cope by working to ensure a better future, and many of them want us to know that, as Anika Jans says, "this is the most essential activity of our times." Because the experience of technology-induced illness leads many people to face the technological crisis in its totality, they perceive that humankind stands at a historical juncture. The ultimatum inherent in this juncture can produce fear and doubt. "I don't feel we are going to survive. Something really serious will happen, probably in the United States," says Diane Carter. "It's hard to imagine a safe world because of the amount of contamination already," says atomic veteran Jose Luis Roybal. "We are headed for destruction."

There is also measured hope. Jane Woolf, whose water supply was contaminated by benzene, feels some hope. "There could be a time when people use technologies in beneficial ways," she says. "In a thousand years consciousness could be completely altered. People used to believe in slavery, and now they don't. I can imagine progress toward a better world in the distant future. But not now. Industry and money are too strong now, and they are what's in control." There is also unbounded hope. "People want change so badly, it *has* to happen," asserts June Casey. "We can do anything we think we can do," adds asbestos worker Loran Calvert.

In the midst of uncertainty about the outcome, either for themselves or for humanity, many technology survivors realize that in order to infuse their lives with meaning they must choose life. They must choose to work for a better future. "To give up," says Andy Hawkinson, "would be unfair to myself, to my family, to the rest of the world." They also understand that the choice is not theirs alone; it must be made by the citizens of technological society. And in their minds, the value of such a choice is bolstered by the realization that the technological predicament was created by humans and can be changed by humans. A crucial aspect of choosing life is to create a tangible vision of a better world.

"The people of the Marshall Islands don't even have a word in their vocabulary for 'enemy,'" says June Casey. "There have been whole peoples who never warred. My vision is of the people leading and the leaders following. I love the people who say 'No, I'm not going to be stopped. No matter how dangerous, I'm going to speak out.' My idea is a beautiful, peaceful society with all the nations of the world working to decrease military spending so those trillions of dollars can be spent for the hungry. My idea is a peaceful revolution."

"So many problems are preventable," says Pat Cody. "I see a world with decent housing, enough food, no tooth-and-claw struggle like we have now. I see happier people who aren't going to inflict emotional injury on one another. No war. No need for all this high technology. And scientific investigation would be carefully studied over a long period of time."

"The only way to survive," asserts Thérèse Khalsa, "is to go back to a culture in which everything is completely natural. Things can be done in a normal, natural, slower way. We don't need all this technology the American life-style makes you believe we need. It's educational to realize that, with most of it, we've only had it for a generation or less. I think people would be happier without it. And a better society would use knowledge, yes, but knowledge that has been proven over thousands of thousands of years—not just in a laboratory test, not just these temporary things that divert everyone."

"I see a smaller society," says dioxin-exposed Gay Ducey, "a human-size one. I see human life and life in general more important than aggression and acquisition, simple approaches to problems accepted and tried."

"I want caring in society," says Irene Baca. "Clean workplaces, clean air and water."

"I have visions of people being empowered," offers Bliss Bruen. "Especially young people who have more to lose. We need global voting, and each vote should be weighted according to how many more years a person has left to live. The younger you are, the more vote you get."

"My five-year-old grandson refuses to eat eggs because they are baby chicks," says Charlotte Mock. "Kids love life. Maybe that's where the hope is."

Technology survivors want, with all their hearts, to look forward with hope and caring, and they want us to join them. They want us to help stop technological destruction and work toward a safe, life-affirming future. They want *us* to choose life.

"My own exposure is just a small piece of the whole," says Marmika Paskiewicz, who was exposed to toxic chemicals. "You hear about people in Pittsburgh cleaning out vats of trichloroethylene. I mean, killing themselves! Three Mile Island! Times Beach! Bedsheets! You can't sleep on cotton bedsheets anymore because they exude formaldehyde! . . . We have to come to a new consciousness, and we need organization on a grass-roots level and throughout the establishment. There aren't any easy answers. *But right now our planet is dying. . . .* I think about this all the time. I try to talk to people, to say something that will be meaningful to them, but so many people have an invisible shield that keeps them from absorbing substantive information about the danger. But we survivors, we have a special mission to let them know. I'd love to see some global organizing. That's what I want to tell everyone. Find out how toxic technologies are affecting you. Then organize to stop it and to find better alternatives. Then reach out to communities in other places with the same problem. Organize!"

EPILOGUE
A UNION OF
TECHNOLOGY
SURVIVORS

We are in the break-the-spell stage.

—**Sarah Pirtle,** DES daughter

Sarah Pirtle delineates a series of stages technology survivors pass through. The first stage is the survival stage: raw coping, keeping up with the demands of illness, getting immediate needs met. Next comes emotional response to the situation: disbelief, anger, fear, grief. The third stage is acceptance, a time for rearranging one's life to accommodate the new situation, reorganizing finances, medical care, psychological support, and legal redress. Fourth is the need for rest. It is a time for emotional distancing and spiritual sustenance. Then comes the assessment stage: joining a survivors' group, gathering information about the offending technology and its medical effects, meeting other similarly affected people, realizing the connection between one's personal experience and the global predicament. Last is what Sarah calls the "break-the-spell" stage: "being powerful in stopping the blight from going further."

Just as with any model of stages, not every person progresses through all the stages or progresses through them in precisely the same order. In fact, given the nature of many technology-induced illnesses, some survivors endure all of Sarah's stages simultaneously. The model is useful, though, because it outlines the basic resources survivors need. They need medical, psychological, spiritual, legal, and political resources. They need meaning in their lives, a sense of connection to forces and issues greater than themselves. Although each survivor has a unique experience, their stories together reveal a pattern that, sad to say, is becoming one of the archetypal human experiences of our times.

In 1987, I coordinated a panel at the Association for Humanistic Psychology's annual meeting in Oakland, California. It was called "Environmental Hazards and Health: Victims Becoming Heroes." I invited Andy Hawkinson, Gay Ducey, Loran Calvert, and Wendy Grace to be present. Each approached this public testimony with trepidation. "I've never told my story in public," confessed Gay. "What could an atomic vet, a naval shipyard worker, an IUD user, and a dioxin-exposed mother have in common?" wondered Andy. "Is anybody interested?" pleaded Wendy. "Does anybody care?"

The obvious result of the panel was that we educated our audience about technological hazards. Some had endured technology-induced illnesses but had never approached them in the context of a widespread problem. Others felt naive about technology and received a new social perspective. All were moved by the testimonies of Andy, Gay, Loran, and Wendy.

The big surprise was the sense of personal connection that arose among the four panelists. Having interviewed them all, I already knew their experiences and feelings were akin. I knew their stories were the same. They didn't know this, and finding out proved to be an emotional turning point. Suddenly, Andy, Gay, Loran, and Wendy discovered that they were not alone. They saw that their immediate community extended far beyond an isolated group affected by one particular technology, and they felt more accepted, connected, and empowered. Afterward, all four told me they would jump at the chance to speak on another panel in the future.

The time for an organization giving technology survivors this chance is now. Having conducted a study of technology survivors' life experiences and psychological responses, I conclude that survivors have important insights to voice about current technological development and human society. They also need support for their own survival. A Union of Technology Survivors—organized by and for survivors—could serve in ways that organizations addressing a single technological threat cannot. By drawing together the population of people made sick by many dangerous technologies, it can cut through the sense of powerlessness that exists even in single-technology organizations. By generalizing the technological threat to a broad social issue, it can break the isolation so many survivors feel. By offering a global perspective on the threat, it can give survivors a context for understanding their experiences and a more powerful foundation from which to educate the public about them.

More specifically, a Union of Technology Survivors could work to:

- Give survivors a forum for telling their stories, giving and receiving support, and securing their identities as loving and lovable human beings.

- Enhance survivors' sense of mission and their public voice for expressing it.

- Organize medical resources for survivors.

- Organize legal resources.

- Organize educational resources.

- Build a base of political power for lobbying and campaigning for survivors' needs.

- Educate the public about health-threatening technologies that are affecting or could affect them.

The research I conducted for this book leads me to reiterate Marmika Paskiewicz's call for "global organizing." We need an international Union of Technology Survivors. This is a call for all technology survivors—from Bhopal residents and Chernobyl workers, to DES children and chemical workers—to look to each other for support and to press us all to see that current trends in technological use cannot continue. This is a call for medical, legal, and educational professionals to share their knowledge and lend a hand. This is a call to break the spell, stop the wounding, and begin the caring.

RESOURCES

ORGANIZATIONS TO CONTACT

SELF-HELP AND GOD'S HELP: MEDICAL AND SPIRITUAL RESOURCES

- **American Academy of Environmental Medicine**
P.O. Box 16106
Denver, CO 80216

  Provides public education on clinical ecology and referrals of qualified physicians; areas of specialty: environmental illness, allergies, immune dysfunction

- **American Association of Naturopathic Physicians**
P.O. Box 33046
Portland, OR 97233
503-255-4863

  Provides education and referrals to naturopathic physicians; lobbies for recognition of naturopathy

- **American Holistic Medical Association and Foundation**
2002 Eastlake Avenue E
Seattle, WA 98102
206-322-6842

  Offers education, research, and national directory to holistic health physicians

- **Antibody Assay Laboratory**
David Roth
1315 Milvia Street
Berkeley, CA 94709
415-525-3169

Provides testing for immune system damage from chemical exposure; can be ordered through any medical doctor, osteopath, or dentist.

- **Association for Holistic Health**
  P.O. Box 12407
  La Jolla, CA 92037
  619-535-0101

  Offers information about holistic health modalities from homeopathy to acupuncture; provides directory of qualified practitioners

- **Association for Humanistic Psychology**
  329 Ninth Street
  San Francisco, CA 94102
  415-626-2375

  Offers education on health, spirituality, and psychology; provides directory of humanistic psychotherapy practitioners

- **Foundation for Advancements in Science and Education**
  Park Mile Plaza
  4801 Wilshire Boulevard
  Los Angeles, CA 90010
  213-937-9911

  Researches, publishes, and holds symposia on the medical aspects of toxification and detoxification by hazardous substances

- **Health Hazard Evaluation Group**
  National Institute for Occupational Safety and Health
  4676 Columbia Parkway
  Cincinnati, OH 45226
  513-841-4382

  Conducts hazard evaluations in workplace and individual medical tests for workers who suspect toxic exposure

- **Holistic Dental Association**
  974 North Twenty-first Street
  Newark, OH 43055
  614-366-3309

  Provides information on practice of toxin-free dentistry and referrals to holistic dentists

- **Huxley Institute for Biosocial Research**
  900 North Federal Highway
  Boca Raton, FL 33432
  800-847-3802
  407-393-6167

  Offers public education, research, medical trainings for professionals, and referrals of qualified physicians utilizing biochemical/orthomolecular medicine; areas of specialty: degenerative diseases, candidiasis, schizophrenia, learning disabilities

- **Interhelp**
  P.O. Box 8895
  Madison, WI 53708
  608-255-1479

  Through international and regional gatherings, support groups, and newsletter, offers spiritual sustenance to people engaged in social change action

- **National Center for Homeopathy**
  1500 Massachusetts Avenue NW—41
  Washington, DC 20005
  202-223-6182

  Provides public education, research library, and directory of homeopathic practitioners

- **National Hospice Organization**
  301 Maple Avenue West—506
  Vienna, VA 22180
  703-938-4449

  Offers public education about the hospice movement and referrals to hospices

- **San Francisco AIDS Foundation**
  333 Valencia Street—4th Floor
  San Francisco, CA 94103
  800-342-AIDS

  Offers AIDS hot line for information

- **Share, Care, & Prayer Ministeries**
  905 North First Avenue
  Arcadia, CA 91506

  Gives spiritual, emotional, and material support to the
  environmentally ill

- **Silent Unity**
  800-821-2935
  816-251-2100

  Offers telephone contact and prayer for people in distress

HELPING OTHERS:
SURVIVOR SUPPORT GROUPS

- *Allergy Action Review*
  7345 Healdsburg Avenue—511
  Sebastopol, CA 95472
  800-824-7163 in U.S.A.
  800-222-9090 in California

  Newsletter to inform and empower environmentally ill people

- **Asbestos Victims of America**
  P.O. Box 559
  Capitola, CA 95010
  408-476-3646

  Assists asbestos survivors with the medical, legal, and psychological
  stresses caused by asbestos exposure; lobbies and engages in legal
  action for compensation

- **Aspartame Consumer Safety Network**
  P.O. Box 19224
  Washington, D.C. 20036
  202-462-8802
  P.O. Box 780634
  Dallas, TX 75378
  214-352-4268

  Provides information to aspartame consumers and the general public

- **Aspartame Victims and Their Friends**
  P.O. Box 1424
  Forest Park, GA 30031

  Provides information and networking for people harmed by aspartame products

- *The Chemical Connection*
  RR1 Box 276 A65
  Wimberly, TX 78676

  Provides network of Texans sensitive to chemicals and magazine, *The Chemical Connection.*

- **Citizen Soldier**
  175 Fifth Avenue
  New York, NY 10010
  212-777-3470

  Provides information and legal support to Agent Orange-exposed and other veterans

- **Citizens Against Pesticide Misuse**
  P.O. Box 67068
  Dallas, TX 75367

  Provides support to pesticide survivors: a clearinghouse of health and safety information and education of health care professionals about pesticide effects

- **Committee on Safety and Health Rights**
  1235 Bay Street
  San Francisco, CA 94123

  Counsels and assists workers who have been discriminated against for raising valid questions about the safety of workplaces

- **Dalkon Shield Information Network**
  P.O. Box 53
  Bethlehem, PA 18016
  215-867-6577

  Provides medical and legal information and support to Dalkon Shield survivors and their families

- **DES Action National**
  Long Island Jewish Hospital
  New Hyde Park, NY 11040
  2845 Twenty-fourth Street
  San Francisco, CA 94110
  415-826-5060

  Educates the public and health professionals about DES exposure; conducts and reports research

- **Disabled Artists Network**
  P.O. Box 20781
  New York, NY 10025

  Provides networking for artists seeking information on hazardous art materials and alternative materials and techniques

- **Human Ecology Action League**
  388 East Gunderson Drive
  Chicago, IL 60628
  212-665-6575

  Assists environmentally ill with information and networking

- **International Alliance of Atomic Veterans**
  P.O. Box 32
  Topock, AZ 86436
  602-768-7515

  Lobbies for medical compensation for radiation-exposed; organizes internationally for nuclear disarmament

- **International Dalkon Shield Victims Education**
  577 Pioneer Building
  Seattle, WA 98104
  206-623-4251

  Defends legal rights of Dalkon Shield survivors

- **National Association of Atomic Veterans**
  P.O. Box 707
  Eldon, MO 65026
  314-392-3361

Promotes rights of atomic veterans to quality medical care and fair legal compensation; lobbies for beneficial legislation

- **National Association of Radiation Survivors**
  Box 20749
  Oakland, CA 94620
  415-655-4886

Represents veterans and civilians exposed to radioactive fallout from nuclear weapons mining, production, testing, and wastes through lobbying, litigation, and public education

- **National Coalition for Cancer Survivorship**
  323 Eighth Street SW
  Albuquerque, NM 87102
  505-764-9956

Promotes advocacy of survivors' rights and needs; provides care and support; develops initiatives to address neglected aspects of survivorship through conferences, support groups, and newsletter

- **National Committee for Atomic Bomb Survivors in the United States**
  1765 Sutter Street
  San Francisco, CA 94115
  415-921-5225

Lobbies for medical assistance for survivors of Hiroshima and Nagasaki

- **National Toxics Campaign**
  29 Temple Place
  Boston, MA 02111
  617-482-1477

Supports citizen actions to clean up hazardous waste sites and establish strong state responsibility and right-to-know policies; offers citizens toxics testing lab for accurate tests of soil, air, and water

- **Occupational Health Legal Rights Foundation**
  815 Sixteenth Street NW
  Washington, DC 20006
  202-842-7840

AFL-CIO–affiliated, assists union and nonunion members in obtaining legal representation for occupational illnesses

- **VDT Coalition**
  Labor Occupational Health Program
  2521 Channing Way
  Berkeley, CA 94720
  415-642-5507

  Researches and does public education on health problems caused by office automation, and in particular, video display terminals

- **9 to 5**
  1224 Huron
  Cleveland, OH 44115
  216-566-9308

  Researches and provides information on health hazards to office workers; organizes workers; lobbies for protection

HELPING THE FUTURE:
PUBLIC INFORMATION, LOBBYING,
AND LEGAL ACTION GROUPS

- **Americans for Safe Food Project**
  1501 Sixteenth Street NW
  Washington, DC 20009
  202-332-9110

  Fights for laws disclosing pesticides, drugs, and other chemicals used in foods and banning pesticides and animal drugs; lobbies for national and state laws to support sustainable, organic agriculture

- **The Bio-Integral Resource Center**
  P.O. Box 7414
  Berkeley, CA 94707
  415-524-2567

  Publishes information on and works to implement environmentally sound pest management programs

- **California Public Interest Research Group**
  215 Pennsylvania Avenue SE
  Washington, DC 20003
  202-546-9707

  Lobbies to enhance the quality of life in California on issues of
  pesticide and toxic reduction, oil drilling, and hazardous waste
  cleanup, and nuclear power liability

- **Center for Science in the Public Interest**
  1501 Sixteenth Street NW
  Washington, DC 20036
  202-332-9110

  Promotes health and nutrition through consumer activism, research,
  and public education on issues of food additives, contaminants to
  water safety

- **Citizens' Clearinghouse for Hazardous Wastes**
  P.O. Box 926
  Arlington, VA 22216
  703-276-7070

  Collects information about toxics; educates the public about toxic
  hazards; assists communities fighting for their rights against toxic
  polluters

- **Earth Island Institute**
  300 Broadway—28
  San Francisco CA, 94133
  415-788-3666

  Sponsors conferences and projects addressing appropriate
  technologies, earth stewardship, and international environmental
  issues; researches, encourages political action

- **Elmwood Institute**
  P.O. Box 5805
  Berkeley, CA 94705
  415-845-4595

  Catalyzes new ways of thinking about human values, technology, and
  culture through symposia, books, films

- **Environ**
  P.O. Box 2204
  Fort Collins, CO 80522
  303-244-0083

  Informs citizens about choices and actions for preventing environmental problems

- **Environmental Action**
  1525 New Hampshire Avenue NW
  Washington, DC 20036
  202-745-4870

  Promotes grass-roots activism to initiate change in areas including pollution, energy resources, and corporate accountability; conducts research; educates the public

- **Environmental Defense Fund**
  257 Park Avenue South
  New York, NY 10010
  212-505-2100

  Seeks solutions to environmental problems through scientific research, legal advocacy, and technical assistance to citizens groups, legislative committees, universities, etc.

- **Environmental Law Institute**
  1616 P Street NW
  Washington, DC 20036
  202-328-5150

  Researches and provides legal assistance on environmental law and policy

- **Greenpeace U.S.A.**
  1436 U Street NW
  Washington, DC 20007
  202-462-1177

  Uses nonviolent tactics to confront institutions damaging the environment by dumping toxics, deploying nuclear weapons, and slaughtering endangered animals

- **Integrated Circuit**
  621 East Nineteenth Street
  Oakland, CA 94606
  415-348-7303

  Links labor and community members who share concern about the
  social impact of new technologies; provides technical assistance to
  workers and community organizations for alternative approaches

- **National Clearinghouse,**
  Committees of Correspondence
  P.O. Box 30208
  Kansas City, MO 64112
  816-931-9366

  Organizes locally based, nationally coordinated programs and actions
  to address ecological crisis from "Green" perspective

- **National Coalition Against the Misuse of Pesticides**
  530 Seventh Street SE
  Washington, DC 20003
  202-543-5450

  Provides information on pesticide use and safe alternatives

- **National Women's Health Network**
  1325 G Street NW
  Washington, DC 20005
  202-347-1140

  Provides information on women's health issues; lobbies to change
  public policy and law regarding women's health

- **Natural Resources Defense Council**
  1350 New York Avenue NW
  Washington, DC 20005
  202-783-7800

  Lobbies for improved environmental protection laws; educates public
  about technological hazards; litigates on EPA implementation

- **The Natural Rights Center**
  P.O. Box 90
  Summertown, TN 38483
  615-964-3992

  Promotes citizen education and activism on issues of technological hazards

- **Northwest Coalition for Alternatives to Pesticides**
  P.O. Box 1393
  Eugene, OR 97440
  503-344-5044

  Works to reduce abuse of pesticides in the Northwest through education and legal action

- **Pacific Studies Center**
  222B View Street
  Mountain View, CA 94041
  415-969-1545

  Provides public interest information about the impact of high-tech electronics

- **Pesticide Action Network International**
  Pesticide Education and Action Project
  P.O. Box 610
  San Francisco, CA 94101
  415-541-9140

  Links groups fighting pesticide misuse in developing countries with similar groups in industrialized nations; educates the public about the health hazards of pesticides

- **Public Citizen**
  2000 P Street NW
  Washington, DC 20036
  202-293-9142

  Defends consumer and worker rights for health and safety through lobbying and litigation; conducts research into safe alternatives to a variety of toxic technologies

- **Sierra Club**
  730 Polk Street
  San Francisco, CA 94109
  415-776-2211

  Addresses environmental crises including population growth, toxic
  pollution, and nuclear waste through public education, lobbying, and
  conservation campaigns

- **Southwest Research and Information Center**
  P.O. Box 4524
  Albuquerque, NM 87106
  505-262-1862

  Works to protect land and water resources in the Southwest by
  researching and providing information and technical assistance to the
  public

- **Texas Center for Policy Studies**
  P.O. Box 2618
  Austin, TX 78768
  512-474-0811

  Promotes ecological practices in Texas through research, education,
  survivor assistance, and public policy

- **The Toxics Coordinating Project**
  2609 Capitol Avenue
  Sacramento, CA 95816
  916-441-4077

  Researches toxics policies in order to prevent problems before they
  happen

- **United Farm Workers of America**
  La Paz
  Keene, CA 93570
  805-822-5571

  Organizes campaigns to ban the use of pesticides; provides survivor
  assistance

- **Women's Occupational Health Resource Center**
  600 West 168th Street
  New York, NY 10032

  Provides information on health risks in the workplace

- **Worldwatch Institute**
  1776 Massachusetts Avenue NW
  Washington, DC 20036

  Researches issues of technological excess from a global perspective; educates the public and elected officials

NOTES

FROM IUDS TO ATOMIC BOMBS

1. "Veterans' Claims for Disabilities from Nuclear Weapons Testing," Hearing before the Committee on Veterans' Affairs (Washington, D.C., U.S. Senate, 1979); and statements by Donald Kerr, acting assistant secretary of defense programs, Department of Energy, at U.S. Congress, House Committee on Interstate and Foreign Commerce, Subcommittee on Health and the Environment, *Effect of Radiation on Human Health* (Washington, D.C.: 95th Cong., 2nd Sess. January 26, 1978), pp. 331–404. The U.S. Department of Defense estimates approximately 210,000 atomic test servicepersons. The National Association of Atomic Veterans calculates the figure between 250,000 and 400,000.

2. *Statistical Abstract of the United States 1974* (Washington, D.C.: U.S. Department of Commerce, Bureau of Census), p. 12. In 1978, 97 percent of individuals tested in Michigan had detectable PBB in their fatty tissue, according to M. Wolff, et al., *Journal of the American Medical Association,* Vol. 247 (1982), p. 2112.

3. W. Pratt and C. Bachrach, *After the Pill, What? National Survey of Family Growth* (Baltimore: National Center for Health Statistics, 1985).

4. Paul Brodeur, *Outrageous Misconduct: The Asbestos Industry on Trial* (New York: Pantheon, 1985), p. 6.

5. Ibid., p. 123.

6. Wasserman and Solomon, *Killing Our Own,* p. 141.

7. "Underground Toxics Leak in Great Lakes and Threaten Residents," *Ground Water Monitor,* 951 Pershing Drive, Silver Springs, Md. 20910 (December 24, 1985).

8. "Pesticide Maker Will Halt Chlordane Sales," *San Francisco Chronicle* (October 1987); and "EPA's Handling of Chlordane Demonstrates FIFRA's Flaws," *Public Citizen* (November/December 1987).

9. Robert Wasserstrom and Richard Wiles, *Field Duty: U.S. Farmworkers and Pesticide Safety,* Study 3 (Washington, D.C.: World Resources Institute, 1985), p. 2.

10. Critical Mass Energy Project, *1986 Nuclear Power Safety Report* (Washington, D.C.: Public Citizen, 1986); and D.F. Ford, *Three Mile Island* (New York: Penguin, 1982).

11. *Glowing on the Job: Worker Exposure to Radiation at Nuclear Power Plants* (Washington, DC: Public Citizen Publications, 1988), Ch. 1.

12. Testimony of Robert Becker, M.D., before the House Subcommittee on Water and Power (Washington, D.C.: September 22, 1987); and *1986 Information Please Almanac* (Boston: Houghton Mifflin, 1986), pp. 713–723.

13. "Pesticide Pollution," *Washington Spectator,* Vol. 14, No. 5 (March 1, 1988), pp. 1–4.

14. David Maraniss and Michael Weisskoff, "Corridor of Death Along the Mississippi," *San Francisco Examiner* (January 31, 1988); and Jay Gould, *Quality of Life in American Neighborhoods* (Boulder, Colo.: Westview Press, 1986), pp. 2.117–2.120.

15. "The Job Ahead: Cleaning Up," *Washington Spectator,* Vol. 14, No. 5 (March 1, 1988), pp. 1–4.

16. J. Johnson and Mark Dowie, "Revenge of the DES Son," *Mother Jones* (February/March 1982); Gina Corea, *The Hidden Malpractice: How American Medicine Mistreats Women* (New York: Harper and Row, 1977, 1985), pp. 275, 292; and Nancy Adess, "The DES Issue: Recommendations for Action" (Paper presented to the U.S. Department of Health and Human Services, Washington, D.C., January 1984).

17. Conversation with Tod Ensign, attorney at Citizen Soldier, July 10, 1989; Michael Uhl and Tod Ensign, *G.I. Guinea Pigs* (New York: Playboy Press, 1980), pp. 210–216; and *In re:* Agent Orange Product Liability Litigation, *597 Federal Supplement,* MDL No. 381 (New York: U.S. District Court, September 25, 1984), p. 756.

18. Morton Mintz, *At Any Cost: Corporate Greed, Women and the Dalkon Shield* (New York: Pantheon, 1985), p. 4.

19. Testimony of Senator Orrin Hatch, Congressional Hearings on Radioactive Fallout from Nuclear Testing (Salt Lake City, April 3, 1983).

20. S. Kelly, "Semiconductor Industry: Layoff Update," *Dataquest Research Newsletter* (San Jose, Calif.: November 1985); Federal Bureau of Labor Statistics, BLS Establishment Survey 790, *Employment and Earnings* (Washington, D.C.: Department of Labor, 1986); and Federal Bureau of Labor Statistics, *Employment and Earnings* (Washington, D.C.: Department of Labor, 1985).

21. Lawrence Altman, "Some Who Use VDTs Miscarried, Study Says," *New York Times* (June 5, 1988); Joint Committee on Behavior and Social Sciences, *Video Displays: Work and Vision* (Washington D.C.: National Academy Press, 1983); Vincent Guiliano, "The Mechanization of Office Work," *Scientific American,* Vol. 247, No. 3 (September 1982); pp. 148–165; and Paul Brodeur, "Annals of Radiation III," *New Yorker* (June 26, 1989), pp. 39–68.

22. Wasserman and Solomon, *Killing Our Own,* p. 131.

23. *Preliminary Assessment of Cleanup Costs for National Hazardous Waste Problems,* Consultant report to the Environmental Protection Agency (Washington, D.C.: Office of Solid Waste, 1979), p. 24; "EPA: 1.2 Million May Be Exposed to Toxic Waste," *Washington Post* (June 6, 1980); and Michael Edelstein, *Contaminated Communities: The Social and Psychological Impacts of Residential Toxic Exposure* (Boulder, Colo.: Westview Press, 1988), p. 3.

24. *Estimates of the Fraction of Cancer Incidence in the United States Attributable to Occupational Factors,* Draft summary (Washington, D.C.: National Institute for Occupational Safety and Health), pp. 2–4.

25. *Ibid.*

26. *Aerometric Information and Retrieval System: 1988* with *Supplemental Data from Regional Office Review* (Washington, D.C.: Environmental Protection Agency, July 1989).

27. United States Environmental Protection Agency, *Unfinished Business: A Comparative Assessment of Environmental Problems* (Washington, D.C.:

U.S. EPA/Office of Policy Analysis (February 1987), pp. 84–86; Lawrie Mott and Karen Snyder, "Pesticide Alert," *Amicus Journal,* Vol. 10, No. 2 (Spring 1988), p. 22; and *Information Please Almanac 1986* (Boston: Houghton Mifflin, 1986), p. 129.

28. Lewis Mumford, *My Works and Days: A Personal Chronicle* (New York: Harcourt Brace Jovanovich, 1979), p. 14.

29. See Appendix for a more complete description of the study.

30. David Noble, *America by Design: Science, Technology and the Rise of the Corporate State* (New York: Alfred Knopf, 1977), p. xxvi.

31. Langdon Winner, *Autonomous Technology: Technics-out-of-Control as a Theme in Political Thought* (Cambridge, Mass.: MIT Press, 1977), p. 8.

32. Lewis Mumford, *Technics and Civilization* (New York: Harcourt Brace and World, 1963), p. 12.

33. Winner, *Autonomous Technology,* p. 12.

34. Lewis Mumford, *The City in History* (New York: Harcourt Brace and World, 1961), pp. 21, 33–34; Noble, *America by Design,* Chs. 8–9.

35. See Edelstein, *Contaminated Communities;* and Henry Vyner, *Invisible Contamination: The Psychosocial Effects of the Invisible Environmental Contaminants* (Lexington, Mass.: Lexington Books, 1988).

36. Ronnie Bulman-Janoff, "The Aftermath of Victimization: Rebuilding Shattered Assumptions," in Charles Figley, ed., *Trauma and Its Wake* (New York: Brunner/Mazel, 1985), p. 16.

THE STORIES

1. Orville Kelly died of malignant lymphoma in 1980 at the age of forty-nine.

2. Cited in *The Progressive,* Vol. 52, No. 3 (March 1988), p. 9; also, Harvey Wasserman and Norman Solomon, *Killing Our Own: The Disaster of America's Experience with Atomic Radiation* (New York: Delta, 1982), pp. 103, 280–281.

3. "Veterans' Claims for Disabilities from Nuclear Weapons Testing," Hearing before the Committee on Veterans' Affairs (Washington, D.C.: U.S. Senate, 1979); and statements by Donald Kerr, acting assistant secretary of defense programs, Department of Energy, at U.S. Congress, House Committee on Interstate and Foreign Commerce, Subcommittee on Health and the Environment, *Effect of Radiation on Human Health* (Washington, D.C.: 95th Cong., 2nd Sess. January 26, 1978), pp. 331–404. The U.S. Department of Defense estimates approximately 210,000 atomic test servicepersons. The National Association of Atomic Veterans calculates the figure between 250,000 and 400,000.

4. Estimates available in the National Association of Radiation Survivors Fact Sheet 1984, Box 20749, Oakland, Calif. 94620.

5. See, for example, the United Nations Science Committee on the Effects of Atomic Radiation, *Sources and Effects of Ionizing Radiation: 1977 Report to the General Assembly* (New York: United Nations, 1977); Advisory Committee on the Biological Effects of Ionizing Radiation, Division of Medical Science, *The Effects on Populations of Exposure to Low Levels of Ionizing Radiation* (Washington, D.C.: National Research Council, National Academy of Sciences, 1972); and John Gofman, *Radiation and Human Health* (San Francisco: Sierra Club Books, 1981).

6. Ernest Sternglass, *Secret Fallout* (New York: McGraw-Hill, 1972), p. 137.

7. Carl Johnson, "Cancer Incidence in an Area of Radioactive Fallout Downwind from the Nevada Test Site," *Journal of the American Medical Association,* Vol. 251, No. 2 (January 13, 1984), pp. 230–236.

8. Jay Truman, "Comparisons Between Chernobyl and Nuclear Testing," *Testing News,* Vol. 4, No. 4 (October 1986), p. 7.

9. Ellen Ruppel Shell, "Sweetness and Health," *Atlantic Monthly* (August 1985).

10. Barbara Mullarkey, "Complaints to FDA Cite Continuing Adverse Reactions to Aspartame," *Wednesday Journal* (January 7, 1987), p. 34; and conversation with Dr. Linda Tollefson, Food and Drug Administration, July 12, 1989.

11. Beatrice Trum Hunter, "Aspartame: The Jury Is Still Out," *Consumer's Research* (January 1986), p. 22.

12. Mallarkey, "Complaints to FDA", p. 34; and Hunter, "Aspartame," p. 23.

13. Hunter, "Aspartame," pp. 23–24.

14. Rif El-Mallakh and Daniel Potenza, "Bittersweet," *Science for the People* (November/December 1988), pp. 17–18.

15. Rene Kimball, "Sawmill Neighbors Air Complaints," *Albuquerque Journal* (November 11, 1986).

16. Committee on Toxicology, National Research Council, *Formaldehyde: An Assessment of Its Health Effects* (Washington, D.C., National Academy Press, 1980); National Research Council, *Formaldehyde and Other Aldehydes* (Washington, D.C.: National Academy Press, 1981); and Consumer Product Safety Commission, "Ban of Urea-Formaldehyde Foam Insulation," *Federal Register,* Vol. 47 (April 2, 1982), pp. 14366–14419.

17. "Attention Sawmill Residents" (Leaflet of the Southwest Organizing Project, October 1986); and Mike Samuels and Hal Bennett, *Well Body, Well Earth* (San Francisco: Sierra Club Books, 1983), pp. 156–157 and 187–188.

18. Alternative Policy Institute, *Toxics and Minority Communities,* Issue Pac No. 2 (Oakland, Calif.: Center for Third World Organizing, 1986), p. 1.

19. *Siting of Hazardous Wastes Landfills and Their Correlation with the Racial and Economic Status of Surrounding Communities* (Washington, D.C.: General Accounting Office, 1983); and Alternative Policy Institute, *Toxics and Minority Communities,* p. 4.

20. Quoted in Sherry Robinson, "Company Wages Battle to Solve Air, Water Pollution Problems," *Business Outlook/Albuquerque Journal* (July 17, 1989), p. 15.

21. John Lecky, "Problems of Trace Anesthetic Levels," in Frederick Orkiin and Lee Cooperman, eds., *Complications in Anesthesiology* (Philadelphia: J.P. Lippincott, 1983), p. 715.

22. *Ibid.,* p. 716.

23. D. Bruce et al., "Causes of Death Among Anesthesiologists: A Twenty-Year Survey," *Anesthesiology,* Vol. 29, No. 565 (1968).

24. Al Vaisman, "Work in Surgical Theaters and Its Influence on the Health of Anesthesiologists," *Eskp. Khir. Anesteziol.,* Vol. 3 (1967), p. 44.

25. "American Society of Anesthesiologists Ad Hoc Committee on the Effect of Trace Anesthetics on the Health of OR Personnel: A National Study," *Anesthesiology,* No. 41, 1974, p. 321.

26. Charles Geraci, Jr., "Operating Room Pollution: Governmental Perspectives and Guidelines," *Nastresia and Analgesia,* Vol. 56 (1977), pp. 775–777.

DISCOVERY

1. Marshall McLuhan, *Understanding Media* (New York: Signet Books, 1964), p. ix.

2. Karen Dorn Steele, "In 1949 Hanford Allowed Radioactive Iodine Into Open Air," *Spokesman* (March 6, 1986).

3. Paul Brodeur, *Outrageous Misconduct: The Asbestos Industry on Trial* (New York: Pantheon, 1985), p. 14.

4. K. Lang and G. Lang, *Collective Dynamics* (New York: Crowell, 1961), p. 71.

5. Andrew Baum, Raymond Fleming, and Jerome Singer, "Coping with Victimization by Technological Disaster," *Journal of Social Issues,* Vol. 39, No. 2 (1983), pp. 117–138.

6. Mardi Horowitz, *Stress Response Syndromes* (New York: Jason Aronson, 1976), p. 56.

7. Anthony Wallace, *Tornado in Worcester: An Exploratory Study of Individual and Community Behavior in an Extreme Situation,* Committee on Disaster Studies, Disaster Study No. 3 (Washington, D.C.: National Academy of Sciences/National Research Council, Publication No. 392, 1956).

8. Lois Gibbs and Murray Levine, *Love Canal: My Story* (New York: Grove Press, 1982), p. 33.

LOSS OF HEALTH

1. Quoted in "The Danger Within," aired on ABC News *20/20*, New York, February 4, 1982.

2. Quoted in Peter de Selding, "A Broken Arrow's Dark Legacy," *Nation* (June 25, 1988), p. 890.

3. Quoted in Lonny Shavelson, "Our Children Are Our Canaries," *California Tomorrow* (Fall 1988), p. 27.

4. Abram Petkau, "Effect of 22 Na+ on a Phospholid Membrane," *Health Physics*, Vol. 22 (1972), p. 239; Abram Petkau, "A Radiation Carcinogenesis from a Membrane Perspective," *Acta Physiologica Scandinavia*, Suppl. Vol. 492 (1980), pp. 81–90; Charles Waldren, Laura Correll, Marguerite Sognier, and Theodore Puck, "Measurement of Low Levels of X-ray Mutagenesis in Relation to Human Disease," *Proceedings of the National Academy of Sciences*, Vol. 83 (1986), pp. 4839–4843; United Nations Committee on the Effects of Atomic Radiation, *Ionizing Radiation: Levels and Effects* (New York: United Nations Publications), 1972; National Research Council Committee on the Biological Effects of Ionizing Radiation, National Academy of Sciences, *The Effects on Populations of Exposure to Low Levels of Ionizing Radiation* (Washington, D.C.: National Academy Press, 1980); and John Gofman, *Radiation and Human Health* (San Francisco: Sierra Club Books, 1981).

5. Harrison Wellington, *Sowing the Wind* (New York: Grossman, 1972), p. 188.

6. P.R. Metcalf and J.H. Holmes, "EEG, Psychological and Neurological Alterations in Humans with Organophosphorus Exposure," *Annals of New York Academy of Sciences*, Vol. 160, No. 1 (1969), pp. 357–365; and Raymond Singer, "Proving Damages in Toxic Torts," *Trial* (November 1985).

7. J.L. Radomski et al., "Pesticide Concentrations in the Liver, Brain, and Adipose Tissue of Terminal Hospital Patients," *Food and Cosmetics Toxicology*, Vol. 6 (1968), pp. 209–225.

8. K.L. Davis, J.A. Savage, and P.A. Berger, "Possible Organophosphate-Induced Parkinsonism," *Journal of Nervous Mental Disorders,* Vol. 166 (1978), pp. 222–225.

9. D. Nag, G.C. Singh, and S. Senon, "Epilepsy Epidemic Due to Benzahexachlorine," *Tropical and Geographical Medicine,* Vol. 29 (1977), pp. 229–232.

10. V.S. Gumenniyi and L.F. Kach, "Findings on Incidence of Diseases of the Cardiovascular System and Respiratory Organs in Areas with Intense and Limited Use of Pesticides," *Pesticide Abstracts,* Vol 10, No. 77-0802 (1976).

11. T.H. Milby and W.L. Epstein, "Allergic Contact Sensitivity to Malathion," *Archives of Environmental Health,* Vol. 9 (1964), pp. 434–437.

12. Radomski et al., "Pesticide Concentrations," pp. 209–225.

13. Cited in Philip Shabecoff, "Hazard Reported in Apple Chemical," *New York Times* (February 2, 1989).

14. United States Environmental Protection Agency, *Unfinished Business: Comparative Assessment of Environmental Problems* (Washington, D.C.: U.S. EPA/Office of Policy Analysis, February 1987), pp. 84–86; and Lawrie Mott and Karen Snyder, "Pesticide Alert," *Amicus Journal,* Vol. 10, No. 2 (Spring 1988), p. 28.

15. M. Loglen et al., "Role of the Endocrine System in the Action of 2,3,7,8-TCDD on the Thymus," *Toxicology,* Vol. 15 (1980), p. 135; R.D. Hinsdill, D.L. Couch, and R.S. Speirs, "Immunosuppression Induced by Dioxin (TCDD) in Feed," *Journal of Environmental Toxicology,* Vol. 4 (1980), p. 401; and J.G. Vos, "Dioxin Induced Thymic Atrophy and Suppression of Thymus-Dependent Immunity," in *Bradbury Report: Biological Method of Dioxin Action* (Cold Springs, Maine: Cold Springs Harbor Laboratory) pp. 401–410.

16. Sheila Hoar et al., "Agricultural Herbicide Use and Risk of Lymphoma and Soft-Tissue Sarcoma," *Journal of American Medical Association,* Vol. 256, No. 9 (September 5, 1986), pp. 1141–1147; Mary Moses, "Cancer in Humans and Potential Occupational and Environmental Exposure in Pesticides," *AAOHN Journal,* Vol. 37, No. 3 (March 1989); Patricia Breslin et al., "Proportionate Mortality Study of Army and Marine Corps Veterans of the Vietnam War" (Washington, D.C.: Office of Environmental Epidemi-

ology/Department of Veterans Affairs); Han Kang et al., "Soft Tissue Carcinoma in the Military Service in Vietnam," *Journal of the National Cancer Institute* (October 1987); and Leonard Hardell, "Association Between Soft Tissue Sarcoma and Exposure to Phenoxyacetic Acid," *Cancer,* Vol. 62, No. 1 (August 1, 1988).

17. V.S. Byers et al., "Association Between Clinical Symptoms and Lymphocyte Abnormalities in a Population with Chronic Domestic Exposure to Industrial Solvent-Contaminated Domestic Water Supply and a High Incidence of Leukemia," *Cancer Immunology Immunotherapy,* Vol. 27 (1988), pp. 77–81.

18. Stephen Dager et al., "Panic Disorder Precipitated by Exposure to Organic Solvents in the Workplace," *American Journal of Psychiatry,* Vol. 144, No. 8 (August 1987), pp. 1056–1058; U. Flodin, C. Edling, and O. Axelson, "Clinical Studies of Psychoorganic Syndromes Among Workers with Exposure to Solvents," *American Journal of Industrial Medicine,* Vol. 5 (1984), pp. 287–295; and K. Lindstrom, H. Ruhimaki, and K. Hamminen, "Occupational Solvent Exposure and Neuropsychiatric Disorders," *Scandinavian Journal of Work Environmental Health,* Vol. 10 (1984), pp. 321–323.

19. S.W. Lagakos, B.J. Wesson, and M. Zelen, *The Woburn Health Study 1984* (Boston: Department of Biostatistics, Harvard School of Public Health, 1984).

20. C. Schoettlin, "Air Pollution and Asthma Attacks in the Los Angeles Area," *Public Health Reports,* Vol. 76 (1961), p. 545; J. Douglass, "Air Pollution and Respiratory Infections in Children," *British Journal of Preventative Social Medicine,* Vol. 20 (1966), p. 1; D. Coffin, "Effect of Air Pollution on Alteration of Susceptibility to Pulmonary Infection," *Proceedings of the Third Annual Conference on Atmospheric Contaminants in Confined Spaces* (1968), p. 75; W. Aronow, "Effect of Freeway Travel on Angina Pectoris," *Annals of Internal Medicine,* Vol. 77 (1972), p. 669; and E. Anderson, "Effect of Low Level Carbon Monoxide Exposure on Onset and Duration of Angina Pectoris," *Annals of Internal Medicine,* Vol. 79 (1973), p. 46.

21. Ralph Dougherty et al., "Sperm Density and Toxic Substances: A Potential Key to Environmental Health Hazards" (Unpublished paper), p. 3.

22. Robert Knapp, Danta Picciano, and Cecile Jacobson, "Y-Chromosomal Iondisjunctional in Dibromochloropropane-Exposed Workers," *Mutation Research,* Vol. 64 (1979).

23. R. Adamson and S.M. Seiber, "Chemically Induced Leukemia in Humans," *Environmental Health Perspective,* Vol. 49 (1981), p. 93; P.F. Infante, "Leukemia in Benzene Workers," *Lancet,* Vol. 2 (1977), pp. 76–78; and R.A. Rinsky et al., "Benzene and Leukemia," *New England Journal of Medicine,* Vol. 316 (1987), pp. 1044–1050.

24. J.R. Reigart and C.D. Graber, "Evaluation of the Humoral Response of Children with Low Level Lead Exposure," *Bulletin of Environmental Contamination Toxicology,* Vol. 16, No. 1 (1976), p. 112; J.A. Thomas and W.C. Brogan III, "Some Actions of Lead on the Sperm of the Male Reproductive System," *American Journal of Industrial Medicine,* Vol. 4, (1983), pp. 127–134; and D. Bellinger et al., "Longitudinal Analyses of Prenatal and Postnatal Lead Exposure and Early Cognitive Development," *New England Journal of Medicine,* Vol. 316 (1987), pp. 1037–1043.

25. K.J. Chang et al., "Immunologic Evaluation of Patients with Polychlorinated Biphenyl Poisoning," *Toxicology and Applied Pharmacology,* Vol. 61 (1981), p. 58; D. P. Brown, "Mortality of Workers Exposed to Polychlorinated Biphenyls: An Update," *Archives of Environmental Health,* Vol. 42, No. 6 (November/December 1987); I. Kalina et al., "Mutagenic and Carcinogenic Effects of Polychlorobiphenyls," *Casopis Lekura Ceckych (Praha),* Vol. 127, No. 14 (April 1, 1988), pp. 426–429; and H. Tsuji et al., "Liver Damage and Heptatocellular Carcinoma in Patients with Yusho," *Fukuoka Igaku Zarshi,* Vol. 78, No. 5 (May 1987), p. 343–348.

26. M.C. Peterson, "Immunotoxic Effects of Ozone in Humans" in I.M. Asher, ed., *Inadvertent Modification of the Immune Response,* United States Federal Drug Administration/Occupational Health Agency, Bulletin No. 80-1074 (Washington, D.C.: U.S. Government Printing Office, 1980), pp. 178–182; R.E. Zelac et al., "Inhaled Ozone as a Mutagen," *Environmental Research,* Vol. 4 (1971), p. 262; and R.R. Guerrerro et al., "Mutagenic Effects of Ozone on Human Cells," *Environmental Research,* Vol. 18 (1979), p. 336.

27. J.G. Bekesi et al., "Investigation of the Immunological Effects of Polybrominated Biphenyls in Michigan Farmers," in J.H. Dean and M. Padaranth Singh, eds., *Biological Relevance of Immune Suppression* (New York: Van Nostrand Reinhold, 1981), pp. 119–135.

28. Foundation for the Advancement of Science and Education, "Subtle Effects of Toxics," *FASE Research Bulletin,* Vol. 7, No. 1 (Spring 1988).

29. Committee on Toxicology, National Research Council, *Formaldehyde: An Assessment of Its Health Effects* (Washington, D.C.: National Academy

Press, 1980); J.C. Harris et al., "Toxicology of Urea-Formaldehyde and Polyurethane Foam Insulation," *Journal of the American Medical Association,* Vol. 245, No. 3 (January 1981); National Research Council, *Formaldehyde and Other Aldehydes* (Washington, D.C.: National Academy Press, 1981); R.C. Anderson et al., "Toxicity of Thermal Decomposition Products of Urea-Formaldehyde and Phenol Formaldehyde foams," *Toxicology and Applied Pharmacology,* Vol. 51, No. 9 (1977); and Consumer Product Safety Commission, "Ban of Urea-Formaldehyde Foam Insulation," *Federal Register,* Vol. 47 (April 2, 1982), p. 14366–14419.

31. Irving Selikoff, Jacob Chung, and Cuyler Hammond, Paper delivered at the Conference on the Biological Effects of Asbestos, New York Academy of Sciences (New York: October 1964); and E. Kagan et al., "Immunological Studies of Patients with Asbestosis I," *Clinical and Experimental Immunology,* Vol. 28 (1977), p. 261; and A. Lange, "An Epistemiological Survey of Immunological Abnormalities in Asbestos Workers II," *Environmental Research,* Vol. 22 (1980), p. 176.

32. Ralph Nader, Ronald Brownstein, and John Richards, eds., *Who's Poisoning America? Corporate Polluters and Their Victims in the Chemical Age* (San Francisco: Sierra Club Books, 1981), p. 12; and *Science for the People,* (January/February 1989), entire issue.

33. Linda Lee Davidoff, "Multiple Chemical Sensitivities," *Amicus Journal,* Vol. 11, No. 1 (Winter 1989), p. 15.

34. Gar Smith, "Ebb and Flow," *Earth Island Journal,* (Fall 1987), p. 3.

35. J. Manson and N. Simons, "Influence of Environmental Agents on Male Reproductive Failure," in Vilma Hunt, ed., *Work and the Health of Women* (Boca Raton, Fla.: CRC Press, 1979); Arthur Bloom, ed., *Guidelines for Studies of Human Populations Exposed to Mutagenic and Reproductive Hazards* (New York: March of Dimes Birth Defects Foundation, 1981), p. 98; G.R. Strobino, J. Klein, and Z. Stein, "Chemical and Physical Exposure of Parents: Effects on Human Reproduction in Offspring," *Journal of Early Human Development,* Vol. 1 (1978), p. 371; Wendy Chavkin and Laurie Welch, *Occupational Hazards to Reproduction* (New York: Program in Occupational Health/Montefiore Medical Center, 1980); 9 To 5 National Association of Working Women, "Analysis of VDT Operator Questionnaires" (February 1984), pp. 1–2; "VDT Safety Controversy Continues," *Public Citizen* (August 1986), p. 5; Lawrence Altman, "Some Who Use VDTs Miscarried, Study Says," *New York Times* (June 5, 1988); and Paul

Brodeur, "Annals of Radiation III," *New Yorker* (June 26, 1989), pp. 39–68.

36. Nancy Wertheimer and Ed Leeper, "Electric Wiring Configurations in Childhood Cancer," *American Journal of Epidemiology*, Vol. 109, No. 3 (1978), p. 273; Hanford Life Sciences Symposium, *Biological Effects of Extremely Low Frequency Electromagnetic Fields* (Hanford, Wash.: Technical Information Center, 1978); Testimony of Robert Becker, M.D., Hearings before the Subcommittee on Water and Power, U.S. House of Representatives, Washington, D.C., September 22, 1987; and Paul Brodeur, "Annals of Radiation I and II," *New Yorker* (June 12 and June 19, 1989), pp. 51–88; pp. 47–73.

37. D. Hirsch et al., "Effect of Oral Tetracycline on the Occurrence of Tetracycline-Resistant Strains of Esch. Coli in the Intestinal Tract of Humans, *Antimicrobial Agents and Chemotherapy*, Vol. 4 (1973), pp. 69–71; G.P. Youmans et al., *The Biologic and Clinical Basis of Infectious Diseases* (Philadelphia: W.B. Saunders, 1975); A. Heindahl and C.E. Nord, "Effect of Phenoxymethylpenicillin and Clindamycin on the Oral, Throat, and Faecal Microflora of Man," *Scandinavian Journal of Infectious Diseases*, Vol. 11 (1979), pp. 233–242; and Marc Lappe, *When Antibiotics Fail* (Berkeley, Calif.: North Atlantic Books, 1986), Ch. 5.

38. W. Rinehard and P.T. Piotrow, "OCs—Update on Usage, Safety and Side Effects," *Population Reports*, Series A, Vol. 6 (1979).

39. D.J. Greenblatt and J. Koch-Weser, "Oral Contraceptives and Hypertension," *Obstetrics and Gynecology*, Vol. 44 (1974).

40. F.A. Lyon and M.J. Frisch, "Endometrial Abnormalities Occurring in Young Women on Long-Term Sequential Oral Contraceptives," *Obstetrics and Gynecology*, Vol. 47 (1976).

41. Phyllis Blair, "Immunologic Consequences of Early Exposure of Experimental Rodents to Diethylstilbestrol and Steroid Hormones," in Arthur Herbst and Howard Bern, eds., *Developmental Effects of Diethylstilbestrol in Pregnancy* (New York: Thieme-Stratton, 1981), pp. 167–178; National Center for Health Statistics, "National Health Interview Survey, U.S. 1985," Vital and Health Statistics Series 10, No. 160, DHHS Publication No. (PHS)86-1588, Public Health Service (Washington D.C.: U.S. Government Printing Office, September 1986); Susan Ways et al., "Alterations in Immune Responsiveness in Women Exposed to Diethylstilbestrol in Utero," *Fertility and Sterility*, Vol. 48, No. 2 (August 1987), pp. 193–197;

and Kenneth Noller et al., "Increased Incidence of Autoimmune Disease Among Women Exposed in Utero to Diethylstilbestrol," *Fertility and Sterility,* Vol. 49, No. 6 (June 1988), pp. 1–3.

42. "Exposure of Male Fetus to DES Tied to Later Genital Problems," *Ob/Gyn News,* Vol. 15, No. 7 (March 1, 1980), p. 7; and Henry Adams, "DES Sons," *DES Action Voice,* Vol. 1. No. 4 (1980).

43. Stanley Robboy et al., "Increased Incidence of Cervical and Vaginal Dysplasia in 3980 DES-Exposed Young Women," *Journal of the American Medical Association,* Vol. 252, No. 21 (December 7, 1984), p. 2979.

44. Arthur Herbst, Marian Hubby, Richard Blough, and Freidoen Azizi, "A Comparison of Pregnancy Experience in DES-Exposed and DES-Nonexposed Daughters," *Journal of Reproductive Medicine,* Vol. 24, No. 2 (February 1980), p. 62; and Merle Berger and Donald Goldstein, "Impaired Reproductive Performance in DES-Exposed Women," *Obstetrics and Gynecology,* Vol. 55, No. 1 (January 1980), pp. 25–27.

45. E.R. Greenberg, "Breast Cancer in Mothers Given DES in Pregnancy," *New England Journal of Medicine* Vol. 311, No. 22 (November 29, 1984).

46. Beatrice Trum Hunter, *The Miracle of Safety: Food Additives and Federal Policy* (New York: Scribner's, 1975), p. 124.

47. M.O. Tisdel et al., in G.E. Inglett, ed., *Symposium: Sweeteners* (New York: AVI Publishing, 1974).

48. William Lijinsky and Samuel Epstein, "Nitrosamines as Environmental Carcinogens," *Nature,* Vol. 225, No. 5227 (1970), pp. 21–23; National Academy of Sciences, "Hazards of Nitrate, Nitrite, and Nitrosamines to Man and Livestock," *Accumulation of Nitrate* (Washington, D.C.: National Academy Press, 1972); R.K. Elespuru and William Lijinsky, "The Formation of Carcinogenic Nitroso Compounds From Nitrite," *Food and Cosmetic Toxicology,* Vol. 11, No. 5 (October 1973), pp. 807–816; and Orville Schell, *Modern Meat: Antibiotics, Hormones and the Pharmaceutical Farm.* (New York: Vintage Books, 1978, 1985).

49. Samuel Epstein, "Concern About Hormone Use on Livestock," *Oakland Tribune* (February 1, 1989).

50. Donald Nilsson, "Sources of Allergenic Gums," *Annals of Allergy,* Vol. 18 (May 1966), pp. 518–524.

51. Edmund Finnerty, "Uticaria and Sodium Benzoate," *Cutis,* Vol. 8, No. 5 (November 1971), pp. 484–485.

52. Theron Randolph, *Human Ecology and Susceptibility to the Chemical Environment* (Springfield, Ill.: Charles Thomas, 1962), p. 71.

53. Elmer Fishernan and Gerald Cohen, "Chemical Intolerance to BHA and BHT," *Annals of Allergy,* Vol. 31, No. 3 (March 1973), pp. 126–133.

54. "Plasticizers Getting into Blood," *Chemical and Engineering News* (February 15, 1971); "Plasticizers: New Entry on List of Suspected Contaminants," *Science News,* Vol. 100 (November 13, 1971), p. 324; and Robert Kelsey, "Evidence Mounts Linking Vinyl Chloride and Cancer," *Chemical and Engineering News* (February 18, 1974).

55. M.A. Boillat et al., *Sozial-und Praventivmedizin (Solothurn),* Vol. 31 (1986), pp. 260–262; I. Rosen et al., *Scandinavian Journal of Work Environment and Health,* Vol. 4 (1978), pp. 184–194; A. Kjellberg et al., *Arbete och Halsa,* Vol. 18 (1979), p. 25; R. Lilis et al., *Environmental Research,* Vol. 15 (1978), pp. 133–138; A.M. Theiss and M. Friedheim, *Scandinavian Journal of Work Environment and Health,* Vol. 3 (1976), pp. 203–214; and H. Checkoway et al., *American Industrial Health Association Journal,* Vol. 43 (1982), pp. 164–169.

THE VICTIM

1. Quoted in Kai Erikson, *Everything in Its Path* (New York: Simon and Schuster, 1976), p. 13.

2. Mark Grossman, "Workers' Noses Sniff Out Chemical Leaks," *Business and Society Review,* No. 59 (Fall 1986), pp. 63–66; and Mark Grossman, "Human Canaries," *Public Citizen,* October 1986, pp. 13–16.

3. J.W. Cook et al., "Sex Hormones and Cancer-Producing Compounds," *Nature* (February 11, 1933), pp. 205–206; M.D. Overholser et al., "Ovarian Hormone and Traumatic Stimulation of Monkey's Cervix." *Proceedings of the Society for Experimental Biology and Medicine.* Vol. 30, No. 9, (June 1933); and Gardner et al., "Stimulation of Abnormal Mammary Growth by Large Amounts of Estrogenic Hormone," *Proceedings of the Society for Experimental Biology and Medicine,* Vol. 33 (1935), pp. 148–150.

4. Nancy Adess, president of DES Action National in 1984, "The DES Issue: Recommendations for Action" (Paper presented to the U.S. Department of Health and Human Services Hearing, Washington, D.C., January 1984).

5. Conversation with Jason Serinus, author of *Psychoimmunity and the Healing Process,* August 20, 1988.

6. M. Bard and O. Sangrers, *The Crime Victim's Book* (New York: Basic Books, 1979); R. Janoff-Bulman and I.H. Frieze, "A Theoretical Perspective for Understanding Reactions to Victimization," *Journal of Social Issues,* Vol. 39 (1983), pp. 1–17; R. Janoff-Bulman and L. Lang-Gunn, "Coping with Disease and Accidents: The Role of Self Blame Attributions," in L.Y. Abramson, ed., *Social-Personal Inference in Clinical Psychology* (New York: Guilford, 1985); R.J. Lifton and E. Olson, "Death Imprint in Buffalo Creek," in H.J. Pared, H.L.P. Resnik, L.G. Pared, eds., *Emergency and Disaster Management* (Bowie Md.: Charles Press, 1976); L.S. Perloff, "Perceptions of Vulnerability to Victimization," *Journal of Social Issues,* Vol. 39 (1983), pp. 41–61; and Martha Wolfenstein, *Disaster: A Psychological Essay* (New York: Arno Press, 1957, 1977).

7. A. Burgess and L. Holmstrom, "Rape Trauma Syndrome," *American Journal of Psychiatry,* Vol. 131 (1974), pp. 981–985.

8. Arthur Stinchcombe et al., *Crime and Punishment: Changing Attitudes in America* (San Francisco: Jossey-Bass, 1980).

9. D. Burdick, "Rehabilitation of the Breast Cancer Patient," *Cancer,* Vol. 36 (1975), pp. 645–648.

10. Irving Janis, "Decision-making Under Stress" in L. Goldberg and S. Breznitz, eds., *Handbook of Stress: Theoretical and Clinical Aspects* (New York: The Free Press, 1982), p. 72.

LOSS OF HELP

1. Henry Vyner, *Invisible Trauma: The Psychological Effects of the Invisible Environmental Contaminants* (Lexington, Mass.: Lexington Books, 1987), p. 130.

2. Marc Pilisuk and Susan Parks, *The Healing Web: Social Networks and Human Survival* (Hanover, N.H.: University Press of New England, 1986).

3. Kai Erikson, *Everything In Its Path* (New York: Simon and Schuster, 1976), p. 194.

4. "Workers vs. Chemical Company," aired on *On Assignment,* KNME TV, Albuquerque, N.M., October 7, 1987.

5. Henry Krystal, "Psychoanalytic Views on Human Emotional Damage," in Bessel Van der Kolb, *Post-Traumatic Stress Disorder: Psychological and Biological Sequelae* (Washington, D.C.: American Psychiatric Press, 1984), p. 21.

6. Robert Waelder, "Psychoanalytic Aspects of War and Peace," *Geneva Studies,* Vol. 10, No. 2 (May 1939), p. I-56.

7. Interview with Henry Vyner, August 12, 1981; also, in Vyner, *Invisible Trauma,* pp. 147–166.

8. Henry Krystal, "Psychotherapy and Survivors of Nazi Persecution" in Henry Krystal, ed., *Massive Psychic Trauma* (New York: International Universities Press, 1968).

9. Samuel Epstein, *The Politics of Cancer* (San Francisco: Sierra Club Books, 1978); Harris Coulter, *Divided Legacy: The Conflict Between Homeopathy and the American Medical Association in the 19th and 20th Centuries* (Berkeley, Calif.: North Atlantic Books, 1982); Paul Starr, *The Social Transformation of American Medicine* (New York: Basic Books, 1984); Robert Livingston, *Confessions of a Medical Heretic* (New York: Warner Books, 1980); and *Hoxsey: Quacks Who Cure Cancer?* directed by Ken Ausubel, Realidad Productions, Santa Fe, New Mexico, 1987.

10. Interview with Henry Vyner, August 12, 1981; also in Vyner, *Invisible Trauma,* p. 62.

11. Veterans Administration, *Federal Benefits for Veterans and Dependents* (Washington, D.C.: United States Government Printing Office, 1987), p. i.

12. Andy Hawkinson, "Survey of Attitudes of Atomic Veterans" (Unpublished paper, 1984).

13. Leslie Guevarra, "V.A.'s Radiation Claims Process Called Too Complex," *San Francisco Chronicle* (September 9, 1987).

14. Dennis Robinette, Seymour Jablon, and Thomas Preston, "Studies of Participants in Nuclear Tests, Sept. 1, 1978–Oct. 31, 1984" (Washington, D.C.: National Research Council, 1985).

15. Abram Petkau, "Effect of 22 Na+ on a Phospholid Membrane," *Health Physics,* Vol. 22 (1972), p. 239; Abram Petkau, "A Radiation Carcinogenesis from a Membrane Perspective," *Acta Physiologica Scandanavia,* Suppl. Vol. 492 (1980), pp. 81–90; Charles Waldren et al., "Measurement of Low Levels of X-ray Mutagenesis in Relation to Human Disease," Proceedings of the National Academy of Sciences, Vol. 83 (1986), pp. 4839–4843; and John Gofman, *Radiation and Human Health* (San Francisco: Sierra Club Books, 1981).

16. "Whose Side Are They On?" aired on ABC News *20/20,* New York, October 2, 1987.

17. Richard Beer, "Extensive Tests Show Westgate Water Safe," *Albuquerque Journal* (May 21, 1982).

LOSS OF HEROISM

1. Ernest Becker, *The Denial of Death* (New York: The Free Press, 1973), p. 5.

2. Kai Erickson, *Everything in Its Path* (New York: Simon and Schuster, 1976); Michael Edelstein, "The Sociological and Psychological Impacts of Groundwater Contamination in the Legler Section of Jackson, New Jersey" (Report to the law firm of Kreindler and Kreindler, 1982); Andrew Baum, Raymond Fleming, and Jerome Singer, "Coping with Victimization by Technological Disaster," *Journal of Social Issues,* Vol. 39, No. 2 (1983), pp. 117–138; Charles Wilkinson, "Aftermath of a Disaster: The Collapse of the Hyatt Regency Hotel Skywalk," *American Journal of Psychiatry,* Vol. 140 (1983), pp. 1134–1139; Stephen Couch and J. Stephen Kroll-Smith, "The Chronic Technical Disaster: Towards a Scientific Perspective," *Social Sciences Quarterly,* Vol. 66 (1985), pp. 564–575; and Henry Vyner, *Invisible Trauma: The Psychosocial Effects of the Invisible Environmental Contaminants* (Lexington, Mass.: Lexington Books, 1988), pp. 52, 79, 137.

3. Quoted in "In Our Water," directed by Meg Switzgable, Foresight Films, Brooklyn, N.Y.; aired on *Frontline,* Public Broadcasting Station, New York, May 23, 1983.

UNCERTAINTY

1. Richard Leiberman, "Uncertainty and Ambiguity: Allies in the Nuclear Age" (Paper presented at Society, Self, and Nuclear Conflict conference, Department of Psychiatry, University of California, San Francisco, and Institute on Global Conflict and Cooperation, University of California, San Francisco, October 19–20, 1985).

2. Henry Krystal, "Psychotherapy with Survivors of Nazi Persecution," in Henry Krystal, ed., *Massive Psychic Trauma* (New York: International Universities Press, 1968).

3. Richard Miller, *Under the Cloud: The Decades of Nuclear Testing* (New York: The Free Press, 1987).

4. *Operations Crossroads: Personnel Radiation Exposure Estimates Should Be Improved.* GAO-RCED-86-15 (Washington, D.C.: General Accounting Office, November 1985).

5. Janet Gardner, "Answers At Last?" *Nation* (April 11, 1987), p. 460.

6. Conversation with Paul Scipione, psychologist with the New Jersey State Agent Orange Commission, January 26, 1988.

7. Frank Butrico, quoted on "Downwind," aired on KUTV TV, Salt Lake City, December 17, 1982; also cited in Howard Ball, *Justice Downwind: America's Atomic Testing Program in the 1950's* (New York: Oxford University Press, 1986), pp. 43–44.

8. "Judge Chides U.S. for Actions in Vets Case," *San Francisco Chronicle* (November 29, 1986).

9. Henry Vyner, *Invisible Trauma: The Psychological Effects of the Invisible Environmental Contaminants* (Lexington, Mass.: Lexington Books, 1987), Ch. 9.

10. Michael Gold, "The Radiowave Syndrome," *Science 80,* premier issue (1980), pp. 79–84; "The Flap Over the Zap," *Newsweek* (July 17, 1978), p. 87; Sally Squires, "Danger: FM Radio," *This World/San Francisco Sunday Examiner* (July 7, 1985), pp. 17–18; and "Cancer, Use of Ham Radio Linked," *Albuquerque Journal* (January 3, 1988).

11. Sheila Hoar et al., "Agricultural Herbicide Use and Risk of Lymphoma and Soft Tissue Sarcoma," *Journal of the American Medical Association,* Vol. 256 (1986), pp. 1141–1147; Mary Moses, "Cancer in Humans and Potential Occupational and Environmental Exposure to Pesticides," *AAOHN Journal,* Vol. 37, No. 3 (March 1989); Patricia Breslin et al., "Proportionate Mortality Study of Army and Marine Corps Veterans of the Vietnam War" (Washington, D.C.: Office of Environmental Epidemiology/ Department of Veterans Affairs); Han Kang et al., "Soft Tissue Carcinoma in the Military Service in Vietnam," *Journal of the National Cancer Institute* (October 1987); and Lennart Hardell and Mikael Erikson, "Association Between Soft Tissue Exposure and Exposure to Phenoxyacetic Acids," *Cancer,* Vol. 62, No. 3. (August 1, 1988), p. 652–656.

12. Nancy Wertheimer and Ed Leeper, "Electrical Wiring Configurations and Childhood Cancer," *American Journal of Epidemiology,* Vol. 109, No. 3 (1978), p. 273; Hanford Life Sciences Symposium, *Biological Effects of Extremely Low Frequency Electromagnetic Fields* (Hanford, Wash.: Technical Information Center, 1978); Testimony of Robert Becker, M.D.; Hearings before the Subcommittee on Water and Power, U.S. House of Representatives, Washington, D.C., September 22, 1987; and Paul Brodeur, "Annals of Radiation I and II, *New Yorker* (June 12 and June 19, 1989), pp. 51–88; pp. 47–73.

13. John Dickerson, "Nutrition and Breast Cancer," *Journal of Human Nutrition,* Vol. 33, No. 21 (1979); and "Eating to Avoid Cancer," *New Scientist* (October 11, 1979).

14. Hans Selye, "A Syndrome Produced by Diverse Nocuous Agents," *Nature,* Vol. 138, No. 32 (1936); and R.H. Rahe, "Subjects' Recent Life Changes and Their Near-Future Illness Susceptibility," *Advances in Psychosomatic Medicine,* Vol. 8, No. 2 (1972).

15. Conversation with Paul Scipione, January 26, 1988.

16. Harvey Wasserman and Norman Solomon, *Killing Our Own: The Disaster of America's Experience with Atomic Radiation* (New York: Dell, 1982), pp. 254–259.

17. J.L. Lyon et al., "Cancer Incidence in Mormons and Non-Mormons in Utah," *New England Journal of Medicine,* Vol. 294, No. 3 (1976), pp. 129–133.

18. M.R. Reich, "Toxic Politics: A Comparative Study of Public and Private Responses to Chemical Disasters in the United States, Italy, and Japan" (Ph.D. dissertation, Yale University, 1982).

19. A.G. Levine, *Love Canal: Science, Politics and People* (Lexington, Mass.: Lexington Books, 1982); and M.R. Fowlkes and P.Y. Miller, "Love Canal: The Social Construction of Disaster" (Paper submitted to the Federal Emergency Management Agency, Washington, D.C., October 1982).

20. Jane Kay, "Noe's 'Cancer Cluster,' " *San Francisco Examiner* (December 13, 1988).

21. Henry Vyner, "The Psychological Effects of Ionizing Radiation," *Culture, Medicine and Psychiatry,* Vol. 7 (1983), p. 248.

22. Hans Selye, *Stress in Health and Disease* (Boston: Butterworth, 1976), p. 725.

23. A. Baum et al., "Chronic and Acute Stress Associated with the Three Mile Island Accident and Decontamination" (Paper submitted to the Nuclear Regulatory Commission, Washington, D.C., July 1981).

24. Vyner, *Invisible Trauma,* p. 144.

25. Ibid., p. 102.

26. American Psychiatric Association, *Diagnostic and Statistical Manual of Mental Disorders,* 3rd ed. (Washington, D.C.: American Psychiatric Association, 1980), p. 236.

RESPONSIBILITY

1. Quoted in "In Our Water," directed by Meg Switzgable, Foresight Films, Brooklyn, N.Y.; aired on *Frontline,* Public Broadcasting Station, New York, May 23, 1983.

2. "Technology On Trial," *Focus: A Quarterly Newsletter on Compliance and Management Issues* (Boston: Commonwealth Films, Fall 1988), p. 1.

3. Ben Druss, "NARS Convention," *PSR Newsletter/San Francisco Bay Area* (Fall 1986), p. 7.

4. Leslie Guevarra, "VA's Radiation Claims Process Called Too Complex," *San Francisco Chronicle* (September 9, 1987).

5. American Law Institute, *Restatement of the Law of Torts,* 2nd ed. (Washington, D.C.: American Law Institute, 1965), Sect. 402A.

6. Joseph Page and Mary-Win O'Brien, *Bitter Wages: Ralph Nader's Study Group on Disease and Injury on the Job* (New York: Grossman Publishers, 1973).

7. Richard Kazis, *Fear at Work: Job Blackmail, Labor, and the Environment* (New York: Pilgrim Press, 1982); and Lawrence Friedman and Jack Ladinsky, "Social Change and the Law of Industrial Accidents," *Columbia Law Review,* Vol. 67 (1967), pp. 269–282.

8. Paul Brodeur, *Outrageous Misconduct: The Asbestos Industry on Trial* (New York: Pantheon, 1985), p. 54.

9. Quoted in *The Atomic Cafe,* directed by Kevin Rafferty, Jayne Loader, and Pierce Rafferty, The Archival Project, Inc., New York, 1982.

10. Lew Gurman, "Human Subjects: American's Guinea Pigs," *Science for the People,* Vol. 20, No. 4 (September/October 1988), p. 16.

11. Robert Alvarez and Arjun Makhijani, "Radioactive Waste," *Technology Review* (August/September 1988), pp. 41–52; and Tom Shanker, "Serious Nuclear Problems," *San Francisco Examiner* (December 12, 1988).

12. Stephen Fox, "Toxic Chemicals and Stress: Anatomy of an Out-of-Court Settlement for Women Workers at GTE Lenkurt Electronics Plant" (Ph.D. dissertation, University of New Mexico, May 1988), Ch. 1.

13. Brodeur, *Outrageous Misconduct,* pp. 31, 36.

14. Frederick Schmidt, "The Legal Labyrinth for Victims of Toxic Wastes," *Business and Society Review,* No. 56 (Winter 1986), p. 57.

15. Brodeur, *Outrageous Misconduct,* p. 39.

16. William Maakestad, "States' Attorneys Stalk Corporate Murders," *Business and Society Review,* No. 56 (Winter 1986), p. 23.

17. Fox, "Toxic Chemicals and Stress," p. 188; and conversation with Stephen Fox, May 15, 1988.

18. Arthur Sharplin, "A Look at the Deal," *The AVA Advisor,* Vol. IV, No. 3 (Fall 1986), p. 9.

19. Mark Grossman, "Workers' Noses Sniff Out Chemical Leaks," *Business and Society Review,* No. 59 (Fall 1986), p. 66.

20. Philip Shabecoff, "E.P.A. Dropping Key Program on Toxic Exposure," *New York Times* (November 24, 1987).

21. Julie Edelson, "Superfund Still in the Dumps," *Technology Review* (November/December 1988), pp. 27–32; and "Superfund Superslow, Study Says," *Albuquerque Journal* (September 9, 1989).

22. Robin Epstein, "Danger! Low-Level Radiation," *Nation* (June 25, 1988), p. 891.

23. Conversation with Marc Lappé, Professor of Humanistic Studies, University of Illinois at Chicago, February 23, 1988.

24. Morton Mintz, *At Any Cost: Corporate Greed, Women, and the Dalkon Shield* (New York: Pantheon, 1985), p. 195.

WE DON'T KNOW THE WOUNDS

1. National Academy of Sciences, "Strategies to Determine Needs and Priorities," in *Toxicity Today* (Washington, D.C.: National Academy Press, 1984).

2. Elliot Diringer, "Science Is Anything but Exact on Toxic Risks," *San Francisco Chronicle* (October 16, 1988).

3. "VDT Controversy Continues," *Public Citizen* (August 1986); and Paul Brodeur, "Annals of Radiation III," *New Yorker* (June 26, 1989), pp. 39–68.

4. "Toxic Waste Cleanup Is Expensive," *Wall Street Journal* (May 16, 1985); and Michael Edelstein, *Contaminated Communities: The Social and Psychological Impact of Residential Toxic Exposure* (Boulder, Colo.: Westview Press, 1988), p. 3.

5. Jacques Ellul, *Propaganda: The Formation of Men's Attitudes* (New York: Vintage Books, 1965), pp. 121, 148.

PSYCHOLOGICAL TABOOS

1. See, for instance: Ernest Becker, *The Denial of Death* (New York: The Free Press, 1973); Matthew Erdelyi and Benjamin Goldberg, "Let's Not Sweep Repression Under the Rug" in John Kihlstrom and Frederick Evans, eds., *Functional Disorders of Memory.* (Hillsdale, New Jersey: Lawrence Erlbaum Associates, 1979); Sigmund Freud, "Repression," in J. Strachey, ed., *The Standard Edition of the Complete Psychological Works of Sigmund Freud,* Vol. 15 (London: Hogarth Press, 1915, 1957); and Mardi Horowitz, "Psychological Responses to Serious Life Events," in Shlomo Breznitz, ed., *The Denial of Stress* (New York: International Universities Press, 1983).

2. Daniel Goleman, *Vital Lies, Simple Truths* (New York: Simon and Schuster, 1985), p. 119.

3. Jacques Ellul, *Propaganda: The Formation of Men's Attitudes* (New York: Vintage Books, 1965), p. 62.

4. Ibid., p. 62.

5. Quoted from advertisement by the Association of Reproductive Health Professionals/Ortho Pharmaceuticals, in *Ms.* (January/February 1989), p. 119.

6. Martha Wolfenstein, *Disaster: A Psychological Essay* (New York: Arno Press, 1957, 1977), p. 28.

7. "Workers Fear Building Toxic Despite Scrubdown" *Gazette Telegraph* (February 6, 1986).

8. Morton Mintz, *At Any Cost: Corporate Greed, Women and the Dalkon Shield* (New York: Pantheon, 1985), pp. 77, 151.

9. Allen v. U.S.A., 588 F. Supp. 247, Utah 1984, p. 211.

10. Cited in Henry Vyner, *Invisible Trauma: The Psychosocial Effects of the Invisible Environmental Contaminants* (Lexington, Mass.: Lexington Books, 1988), p. 185.

11. Quoted in Allen v. U.S.A., pp. 289–290.

12. Carl Johnson, "Cancer Incidence in an Area of Radioactive Fallout Downwind from the Nevada Test Site," *Journal of the American Medical Association,* Vol. 251 (January 13, 1984), p. 231; Joseph Lyon et al., "Childhood Leukemia Associated with Fallout from Nuclear Testing," *New England Journal of Medicine,* Vol. 300 (1979), pp. 397–402; Charles Smart and Joseph Lyons, eds., *Cancer in Utah, 1964–1977,* Report No. 3 (Salt Lake City: Utah Cancer Registry, September 1979), p. 1; Transcript of Special Town Meeting conducted by U.S. Senator Orrin Hatch (R-Utah) in St. George, Utah, April 17, 1979; and Edward Weiss, "Leukemia Mortality in Southwestern Utah, 1950–1964" (Report for the Atomic Energy Commission, 1965.

13. Robert Gillette, "Tests Show Slight Harm from Fallout," *Albuquerque Journal* (November 7, 1988).

14. Porsche advertisement, 1987.

15. Sam Keen, *Faces of the Enemy: Reflections of the Hostile Imagination* (San Francisco: Harper and Row, 1986), p. 19.

16. Fritz Perls, *Gestalt Approach and Eye Witness to Therapy* (Ben Lomond, Calif.: Science and Behavior Books, 1973), p. 35.

17. D. P. Brown, "Mortality of Workers Exposed to Polychlorinated Biphenyls: An Update," *Archives of Environmental Health,* Vol. 42, No. 6 (November/December 1987); I. Kalina et al., "Mutagenic and Carcinogenic Effects of Polychlorobiphenyls," *Casopis Lekura Cechych (Praha),* Vol. 127, No. 14 (April 1, 1988), pp. 426–429; Kathleen Kreys et al., "Association of Blood Pressure and Polychlorinated Biphenyl Levels," *Journal of the American Medical Association,* Vol. 245, No. 24 (June 26, 1981), pp. 2505, 2509; W. J. Rogan et al., "Congenital Poisioning by Polychlorinated Biphenyls and Their Contaminants in Taiwan," *Science,* Vol. 241, No. 4863 (July

15, 1988), pp. 334–336; and H. Tsuyi et al., "Liver Damage and Hepatocellular Carcinoma in Patients with Yusho," *Fukuoka Igaku Zasshi,* Vol. 78, No. 5 (May 1987), pp. 343–348.

18. "Workers Fear Building Toxic Despite Scrubdown," *Gazette Telegraph* (February 6, 1986).

19. Carol Wolman, "Current Consciousness About Nuclear Weapons in the United States" (Paper delivered at the Second International Conference on Psychological Stress and Adjustment in Time of War and Peace, Jerusalem, Israel, June 22, 1978).

20. Steven Kull, "Nuclear Nonsense," *Foreign Affairs,* No. 58 (Spring 1985); and *Minds at War* (New York: Basic Books, 1988).

21. Goleman, *Vital Lies,* pp. 95–101.

SOCIAL TABOOS

1. Kate Millpointer and Preston Truman, "Silent Summer," *In These Times* (November 23–December 6, 1988), p. 12–13, 22; "Chernobyl Meltdown," *Newsweek* (May 12, 1986), pp. 20–49; and Ernest Sternglass, "The Implications of Chernobyl for Human Health," *International Journal of Biosocial Research,* Vol. 8, No. 1 (1986), pp. 7–36.

2. Willis Harman, "Peace, Beliefs and Legitimacy," in Mark Macy, ed., *Solutions for a Troubled World* (Boulder, Colo.: Macy Press, 1987), pp. 124–129.

3. General Electric advertisement.

4. Cited in Don Fabun, ed., *The Dynamics of Change* (Englewood Cliffs, N.J.: Prentice-Hall, 1970), p. II-22; and Edward Hall, *The Silent Language* (Greenwich, Conn.: Fawcett, 1965).

5. David Noble, *America By Design: Science, Technology, and the Rise of Corporate Capitalism* (New York: Alfred A. Knopf, 1977), p. xi.

6. Lewis Mumford, *My Works and Days: A Personal Chronicle* (New York: Harcourt Brace Jovanovich, 1979), p. 7.

7. Du Pont advertisement.

8. Lewis Mumford, "The Case Against Modern Architecture," *Architectural Record,* Vol. 131, No. 4 (April 1962), p. 157.

9. Langdon Winner, *Autonomous Technology: Technics-out-of-Control as a Theme in Political Thought* (Cambridge, Mass.: MIT Press, 1977), p. 124.

10. IBM advertisement, 1988.

11. Jerry Mander, *The Least Popular Cause* (San Francisco: Sierra Club Books, forthcoming in 1990).

12. Ibid.

13. Daniel Zwertling on *All Things Considered,* National Public Radio, Washington, D.C., July 9, 1988.

14. Ithiel de Sola Pool, *Forecasting the Telephone* (Norwood, N.J.: ABLEX, 1983).

15. Paul Ceruzzi, "An Unforeseen Revolution: Computers and Expectations," in Joseph Corn, ed., *Imagining Tomorrow: History, Technology, and the American Future* (Cambridge, Mass.: MIT Press, 1986), pp. 188–201.

16. Mander, *Least Popular Cause.*

17. Kanji Kuramoto, Lecture presented at National Radiation Survivors Conference, Berkeley, Calif., August 1986.

18. MasterCard advertisement, 1987.

19. Jacques Ellul, *Technological Society,* trans. John Wilkinson (New York: Alfred A. Knopf, 1964), p. 43.

20. Winner, *Autonomous Technology,* p. 20.

21. Mander, *Least Popular Cause.*

22. Lieutenant Colonel Edward Williams about need for $70 million blast simulator at White Sands Missile Range in New Mexico; quoted in David Morrissey, "White Sands Proposed as Site for Nuclear Blast Simulator," *Albuquerque Journal,* (February 28, 1988).

23. *Cheaper by the Dozen,* directed by Walter Lang, Twentieth Century Fox, 1950.

24. Quoted in Siegfried Giedion, *Mechanization Takes Command* (New York: Norton and Co., 1948), p. 99.

25. *Modern Times,* directed by Charlie Chaplin, United Artists, 1936.

26. Marshall McLuhan, *Understanding Media* (New York: McGraw-Hill, 1964), p. 120.

27. American Airlines advertisement, 1988.

28. Barnett Newman, "The First Man Was an Artist," *The Tiger's Eye,* Vol. 1 (October 1947), p. 59.

29. Winner, *Autonomous Technology,* p. 129.

30. Advertisement for Compaq Deskpro 386/33, 1989.

31. Paul Ceruzzi, "An Unforeseen Revolution," in Joseph Corn, ed., *Imagining Tomorrow,* p. 188.

32. S. Kelly, "Semiconductor Industry: Layoff Update," *Dataquest Research Newsletter* (San Jose, Calif.: November 1985); Federal Bureau of Labor Statistics, BLS Establishment Survey 790, *Employment and Earnings* (Washington, D.C.: Department of Labor, 1986); and Federal Bureau of Labor Statistics, *Employment and Earnings* (Washington, D.C.: Department of Labor, 1985).

33. Quoted in Don Fabun, *The Dynamics of Change* (Englewood Cliffs, N.J.: Prentice-Hall, 1970), p. IV-15.

SELF-HELP

1. Michael Edelstein, *Contaminated Communities: The Social and Psychological Impacts of Residential Toxin Exposure* (Boulder, Colo.: Westview Press, 1988), p. 87.

2. Henry Krystal, "Trauma and Affects," in A. Solnit, ed., *Psychoanalytic Study of the Child* (New Haven: Yale University Press, 1978).

3. Sigmund Freud and R. Brever, *Studies in Hysteria* (Boston: Beacon Press, 1980).

4. Elisabeth Kübler-Ross, Lecture at "Life, Death, and Transitions" Workshop, Valley Center, Calif., August 1978; and Elisabeth Kübler-Ross, *On Death and Dying* (New York: Macmillan, 1969), Chapters 3–7.

5. E. Parson, "The Reparation of the Self: Clinical and Theoretical Dimensions in the Treatment of Vietnam Combat Veterans," *Journal of Contemporary Psychotherapy*, Vol. 14, No. 1 (1984), pp. 4–56.

6. Henry Vyner, "The Psychological Effects of Ionizing Radiation," *Culture, Medicine and Psychiatry*, Vol. 7 (1983), pp. 241–263; and Henry Vyner, *Invisible Trauma: The Psychosocial Effects of the Invisible Environmental Contaminants* (Lexington, Mass.: Lexington Books, 1988), p. 136.

7. Henry Krystal, "Trauma and the Stimulus Barrier." (Unpublished paper presented at the annual meeting of the American Psychoanalytic Association, San Francisco, May 8, 1970); Ronnie Janoff-Bulman, "The Aftermath of Victimization: Rebuilding Shattered Assumptions," in Charles Figley, ed., *Trauma and Its Wake: The Study and Treatment of Post-traumatic Stress Disorder* (New York: Brunner/Mazel, 1985), pp. 15–35; S. Epstein, "The Self and Concept Revisited: Or a Theory of a Theory," *American Psychologist*, Vol. 28 (1973), pp. 404–416; C.M. Parkes, "Psychosocial Transitions: A Field for Study," *Social Science and Medicine*, Vol. 5 (1971), pp. 101–115; and Peter Marris, *Loss and Change* (Garden City, N.J.: Anchor/Doubleday, 1975).

8. L.R. Hubbard, "The Purification Rundown Replaces the Sweat Program," *HCO Bulletin* (December 4, 1979); and H.R. Hubbard, *The Technical Bulletins*, Vol. 12 (Los Angeles: Bridge Publications, 1980), pp. 163–181.

9. David Schnare, Max Ben, and Megan Shields, "Body Burden Reductions of PCBs, PBBs, and Chlorinated Pesticides in Human Subjects," *Ambio*, Vol. 13, Nos. 5–6 (1984), pp. 378–380.

10. Hubbard, *Technical Bulletins*, pp. 163–181; and David Schnare et al., "Evaluation of a Detoxification Regimen for Fat Stored Xenobiotics," *Medical Hypothesis*, Vol. 9 (1982), pp. 265–282.

11. Edelstein, *Contaminated Communities*, p. 6.

12. A complete list of self-help associations for technology survivors is listed in the Resources section at the end of this book.

13. Martha Wolfenstein, *Disaster: A Psychological Essay* (New York: Arno Press, 1957, 1977), p. 82.

GOD'S HELP

1. Report by Henry Vyner in workshop on Psychological Aspects of Radiation Exposure, Radiation Survivors Congress, San Francisco, Calif., October 1984.

2. Jean Houston, *The Search for the Beloved* (Los Angeles: Jeremy Tarcher, 1987), pp. 104–106.

3. Ibid., p. 106.

4. Sigmund Freud, "A Religious Experience," in *Collected Papers*, Vol. 5 (London: Hogarth Press, 1950), pp. 243–246.

5. Martha Wolfenstein, *Disaster: A Psychological Essay* (New York: Arno Press, 1957, 1977), p. 61.

6. F. Haronian, "A Psychosynthesis Model of Personality and Its Implications for Psychotherapy," *Journal of Humanistic Psychology*, Vol. 15 (1975), pp. 25–51.

7. Conversation with Stephen Fox, American Studies scholar, University of New Mexico, Albuquerque, July 13, 1988.

8. Lois Gibbs and Murray Levine, *Love Canal: My Story* (New York: Grove Press, 1982), p. 160.

HELPING OTHERS

1. "Probing a Mysterious Cluster," *Newsweek* (February 23, 1987).

2. Martha Wolfenstein, *Disaster: A Psychological Essay* (New York: Arno Press, 1957, 1977), p. 92.

3. Ibid., p. 98.

4. Marc Pilisuk and Susan Parks, "The Place of Network Analysis in the Study of Supportive Social Associations," *Basic and Applied Psychology,* Vol. 2, No. 2 (1981), pp. 121–135; and Marc Pilisuk and Susan Parks, *The Healing Web: Social Networks and Human Survival* (Hanover, N.H.: University Press of New England, 1986), p. 17.

5. Conversation with Shepherd Bliss, Professor of Psychology, J.F. Kennedy University, Orinda, Calif., November 1985.

HELPING THE FUTURE

1. "Protestor Guilty in Attack on U.S. Computer," *San Francisco Chronicle* (November 17, 1987).

2. "Mowing Down Caltrans," *San Francisco Examiner* (June 3, 1987).

3. "Kicking the TV Habit at U.C.," *San Francisco Chronicle* (November 14, 1987).

4. "Cesar Chavez Ends Grape Boycott Fast," *Albuquerque Journal* (August 22, 1988).

5. Adrian Randall, "The Philosophy of Luddism: The Case of the West of England Wool Workers, ca. 1790–1809," *Technology and Culture,* Vol. 27, No. 1 (January 1986).

6. Quoted from a speech by Mario Savio, sit-in rally, Berkeley, Calif., December 2, 1964. Transcribed from "Is Freedom Academic?" radio documentary, KPFA, Berkeley, Calif.

7. Marc Pilisuk, Susan Parks, and Glenn Hawkes, "Public Assessment of Technological Dangers: Who Cares? Who Acts?" (Study for the California Department of Food and Agriculture and the University of California, Davis, 1986).

8. Irving Janis, *Airwar and Emotional Stress: Psychological Studies of Bombing and Civilian Defense* (New York: McGraw-Hill, 1951); J.W. Powell, *An Introduction to the Natural History of Disaster,* unpublished report (Baltimore, Md.: Psychiatric Institute of the University of Maryland, June

1954); and Eli Marks et al., *Human Reactions in Disaster Situations,* Report No. 52 (Washington, D.C.: National Opinion Research Center, June 1954).

9. Kari Pratt, "Letter to the Editor," *Jefferson Davis County News* (June 1988).

10. See the Resources section for a more complete listing.

11. *New York Times* (March 28, 1982).

THE EARTH IS A SURVIVOR TOO

1. The first actual photograph of the whole earth, a very crude picture, was taken aboard the U.S. satellite *Explorer VI* in August 1959.

2. Amory Lovins and Hunter Lovins, *Brittle Power: Energy Strategy for National Security* (Andover, Mass.: Brick House Publishing, 1982), p. 141.

3. Lester Brown et al., *State of the World 1986* (New York: W.W. Norton, 1986), p. 119.

4. Economic Regulatory Administration, *The National Electricity Reliability Study,* February draft, Vol. I, DOE/RG-0055 (Washington, D.C.: Office of Utility Systems, 1981), pp. 2–9.

5. Lovins and Lovins, *Brittle Power,* p. 124.

6. Gene Likens et al., "Acid Rain," *Scientific American* (October 1979), p. 49; and John Holum, *Topics and Terms in Environmental Problems* (New York: John Wiley and Sons, 1977), p. 619.

7. Lester Brown et al., *State of the World 1985* (New York: W.W. Norton, 1985), p. 15.

8. Ibid., p. 29.

9. Robert Wasserstrom and Richard Wiles, *Field Duty: U.S. Farmworkers and Pesticide Safety,* Study 3 (Washington, D.C.: World Resources Institute, 1985), p. 2.

10. Karin Winegar, "Plastics: The Once and Future Trash," *Utne Reader* (November/December 1987), p. 24.

11. Figures from the International Trade Commission, quoted in Ralph Nader, Ronald Brownstein, and John Richard, *Who's Poisoning America?* (San Francisco: Sierra Club Books, 1981), p. 5.

12. L. Fishbein, *Potential Industrial Carcinogens and Mutagens,* 560/5-77-005 (Washington, D.C.: U.S. Environmental Protection Agency, 1977); and Thomas Maugh II, "Chemical Carcinogens: The Scientific Basis for Regulation," *Science* (September 29, 1978), pp. 1200–1205.

13. Division of Water, "Comparison of 1981–82 and 1985–86 Toxic Substance Discharges to the Niagara River" (New York State: Department of Environmental Conservation, August 1987), p. 3.

14. Forestry Division, *Tropical Forest Resources,* Forestry Paper 30 (Rome, Italy: United Nations Food and Agriculture Organization, 1982); and Gunter Schramm and David Shirah, "Sub-Saharan Africa Policy Paper: Energy" (Washington, D.C.: World Bank, August 20, 1984).

15. Sara Pacher, "The World According to Lester Brown," *Utne Reader* (September/October 1987), p. 87.

16. Conversation with Louis Thompson, Railroad Division, World Bank, Washington, D.C., September 3, 1987.

17. *Highways Statistics 1985* (Washington, D.C.: Federal Highways Administration, 1985), p. 109.

18. Ibid., p. 17.

19. Brown et al., *State of the World 1985,* pp. 106–107.

20. Randolph Harrison, "March out of Poverty Lays Waste to Land," *Orlando Sentinel* (March 10, 1987).

21. Frederick Ordway, *Pictorial Guide to the Planet Earth* (New York: Thomas Crowell, 1975), p. 39.

22. Testimony of Paul Tierrey, Trans-Alaskan Pipeline System Hearings, Federal Energy Regulatory Commission, Washington, D.C., October 30, 1978.

23. William Arkin and Richard Fieldhouse, *Nuclear Battlefield: Global Links in the Arms Race* (Cambridge, Mass.: Ballinger/Harper and Row, 1985), p. 2.

24. Fred Shapiro, "Radwaste in the Indians' Backyards," *Nation* (May 7, 1983), p. 574.

25. Michael Brown, *Laying Waste: The Poisoning of America by Toxic Chemicals* (New York: Pantheon, 1979, 1980), p. 293.

26. Office of Solid Waste, *Everybody's Problem: Hazardous Wastes* (Washington, D.C.: Environmental Protection Agency, 1981), p. 1.

27. Office of Solid Waste, *Preliminary Assessment of Cleanup Costs for National Hazardous Waste Problems,* Consultant Report (Washington, D.C.: Environmental Protection Agency, 1979), p. 24.

28. Arkin and Fieldhouse, *Nuclear Battlefields,* p. 34.

29. "A Primer on Nuclear Testing," *Testing News,* Vol. III, No. IV (August 1985).

30. Ordway, *Pictorial Guide,* p. 29.

31. Lester Brown, *The Twenty Ninth Day* (New York: W.W. Norton, 1978), p. 22.

32. Arkin and Fieldhouse, *Nuclear Battlefields,* p. 44.

33. Ordway, *Pictorial Guide,* pp. 132–133.

34. Richard Hudson, "Atomic Age Dump: A British Nuclear Plant Recycles Much Waste," *Wall Street Journal* (April 11, 1984).

35. Stockholm International Peace Research Institute, *Outer Space: Battlefield of the Future?* (London: Taylor and Francis Ltd., 1978), pp. v, 2.

36. Peter Coy, "Satellites Affect Face of Communications," *Albuquerque Journal* (July 12, 1987).

37. Gar Smith, "Space as a Wilderness," *Earth Island Journal* (Winter 1987), p. 24.

38. Bill Lawren, "Six Scientists Who May Save the World," *Omni* (September 1987), p. 96.

WHAT TECHNOLOGY SURVIVORS
WANT TO TELL US

1. Joan Wescott, "Victims and Visionaries of the Planet," *The Allergy Store Catalog* (Spring 1988), p. 67.

2. Ibid., p. 67.

BIBLIOGRAPHY

TECHNOLOGY, CULTURE, AND HISTORY

Berry, Wendell. *The Unsettling of America.* San Francisco: Sierra Club Books, 1977.

Braverman, Harry. *Labor and Monopoly Capital: The Degradation of Work in the Twentieth Century.* New York: Monthly Review Press, 1974.

Burnham, David. *The Rise of the Computer State.* New York: Random House, 1983.

Commoner, Barry. *The Closing Circle.* New York: Bantam, 1972.

Ellul, Jacques. *The Technological Society.* New York: Vintage Books, 1964.

———. *Propaganda: The Formation of Men's Minds.* New York: Vintage Books, 1965.

Epstein, Samuel. *The Politics of Cancer.* San Francisco: Sierra Club Books, 1978.

Giedion, Siegfried. *Mechanization Takes Command.* New York: Norton, 1948.

Griffin, Susan. *Woman and Nature: The Roaring Inside Her.* New York: Harper and Row, 1978.

Hughes, Thomas Parke. *Changing Attitudes Toward American Technology.* New York: Harper and Row, 1975.

Loeb, Paul. *Nuclear Culture.* Philadelphia: New Society, 1987.

Mander, Jerry. *Four Arguments for the Elimination of Television.* New York: Quill, 1978.

———. *The Least Popular Cause.* San Francisco: Sierra Club Books, 1990.

McLuhan, Marshall, *Understanding Media.* New York: Signet, 1964.

Merchant, Carolyn. *The Death of Nature.* San Francisco: Harper and Row, 1980.

Mumford, Lewis. *The Myth of the Machine,* Vol. 1, *Technics and Human Development.* New York: Harcourt, Brace and World, 1967.

———. *The Myth of the Machine,* Vol. 2, *The Pentagon of Power.* New York: Harcourt, Brace and World, 1970.

———. *Technics and Civilization.* New York: Harcourt, Brace, 1934.

Noble, David. *America by Design: Science, Technology, and the Rise of Corporate Capitalism.* New York: Alfred Knopf, 1977.

Norris, Ruth, ed. *Pills, Pesticides, and Profits: International Trade and*

Toxic Substances. Croton-on-the-Hudson, N.Y.: North River Press, 1982.

Perrucci, Robert, and Marc Pilisuk, eds. *The Triple Revolution.* Boston: Little, Brown, 1971.

Winner, Langdon. *Autonomous Technology: Technics-out-of-Control as a Theme in Political Thought.* Cambridge, Mass.: MIT Press, 1977.

Zerzan, John, and Alice Cannes, eds., *Questioning Technology: A Critical Anthology.* London: Freedom Press, 1988.

TECHNOLOGICAL HAZARDS

Americans for Safe Foods, *Guess Who's Coming to Dinner? Contaminants in Our Food.* Washington, D.C.: Center for Science in the Public Interest, 1987.

Ball, Howard. *Justice Downwind: American's Atomic Testing Program in the 1950s.* New York: Oxford University Press, 1986.

Becker, Robert, and Gary Selden. *The Body Electric.* New York: Quill, 1985.

Berman, Daniel. *Death on the Job: Occupational Health and Safety Struggles in the United States.* New York: Monthly Review Press, 1978.

Brodeur, Paul. *Outrageous Misconduct: The Asbestos Industry on Trial.* New York: Pantheon, 1985.

———. *The Zapping of America: Microwaves, Their Deadly Risk, and the Cover Up.* New York: Norton, 1977.

Brown, Michael. *Laying Waste: The Poisoning of America by Toxic Chemicals.* New York: Pantheon, 1980.

———. *The Toxic Cloud: The Poisoning of America's Air.* New York: Perennial, 1987.

Carson, Rachel. *The Silent Spring.* Boston: Houghton Mifflin, 1962.

Corea, Gena. *The Hidden Malpractice: How American Medicine Mistreats Women.* New York: Harper and Row, 1977, 1985.

Coyle, Dana, et al., *Deadly Defense: Military Radioactive Landfills.* New York: Radioactive Waste Campaign, 1988.

Epstein, Samuel, Lester Brown, and Carl Pope. *Hazardous Waste in America.* San Francisco: Sierra Club Books, 1982.

Franklin, Melia. *Toxics and Minority Communities,* Issue Pac No. 2. Oakland, Calif.: Alternative Policy Institute of the Center for Third World Organizing, July 1986.

Gofman, John. *Radiation and Human Health.* San Francisco: Sierra Club Books, 1981.

Lappé, Marc. *When Antibiotics Fail.* Berkeley, Calif.: North Atlantic Books, 1986.

Miller, Richard. *Under the Cloud: The Decades of Nuclear Testing.* New York: The Free Press, 1987.

Mintz, Morton. *At Any Cost: Corporate Greed, Women, and the Dalkon Shield.* New York: Pantheon, 1985.

Nader, Ralph, Ronald Brownstein, and John Richard, eds. *Who's Poisoning America: Corporate Polluters and Their Victims.* San Francisco: Sierra Club Books, 1981.

National Research Council Committee on the Biological Effects of Ionizing Radiation. *The Effects on Populations of Exposure to Low Levels of Ionizing Radiation.* Washington, D.C.: National Academy Press, 1980.

Page, J.A. *Bitter Wages: Ralph Nader's Study Group Report on Disease and Injury on the Job.* New York: Grossman Press, 1973.

Perrow, Charles. *Normal Accidents: Living with High-Risk Technology.* New York: Basic Books, 1984.

Piller, Charles, and Keith Yamamoto. *Gene Wars: Military Control Over the New Genetic Technologies.* New York: Beech Tree/William Morrow, 1988.

Powledge, Fred. *Water: The Nature, Uses, and Future of our Most Precious and Abused Resource.* New York: Farrar, Straus, and Giroux, 1982.

Schell, Orville. *Modern Meat: Antibiotics, Hormones, and the Pharmaceutical Farm.* New York: Vintage, 1985.

Siegel Lenny, and John Markoff. *The High Cost of High Tech: The Dark Side of the Chip.* New York: Harper and Row, 1986.

Sternglass, Ernest. *Secret Fallout: Low-Level Radiation from Hiroshima to Three Mile Island.* New York: McGraw-Hill, 1972, 1981.

Uhl, Michael, and Ted Ensign. *GI Guinea Pigs.* New York: Playboy Press, 1980.

Van Strum, Carol. *A Bitter Fog: Herbicides and Human Rights.* San Francisco: Sierra Club Books, 1983.

Wasserman, Harvey, and Norman Solomon. *Killing Our Own: The Disaster of America's Experience with Atomic Radiation.* New York: Dell, 1982.

Wasserstrom, Robert, and Richard Wiles. *Field Duty: U.S. Farmworkers and Pesticide Safety.* Washington, D.C.: World Resources Institute, 1985.

Weir, David, and Mark Shapiro. *Circle of Poison: Pesticides and People in a Hungry World.* San Francisco: Institute for Food and Development Policy, 1981.

Whiteside, Thomas. *The Pendulum and the Toxic Cloud.* New Haven, Conn.: Yale University Press, 1979.

THE PSYCHOSOCIAL IMPACT OF TECHNOLOGY-INDUCED DISASTER AND DISEASE

Brod, Craig, and Wes St. John. *Technostress: The Human Cost of the Computer Revolution.* Reading, Mass.: Addison-Wesley, 1986.

Edelstein, Michael. *Contaminated Communities: The Social and Psychologi-*

cal Impacts of Residential Toxic Exposure. Boulder, Colo.: Westview, 1988.

Erikson, Kai. *Everything in Its Path.* New York: Simon and Schuster, 1976.

Gibbs, Lois. *Love Canal: My Story.* New York: Grove Press, 1982.

Glendinning, Chellis. *Waking Up in the Nuclear Age.* New York: Beech Tree/William Morrow, 1987.

Grosser, George, et al., eds. *The Threat of Impending Disaster.* Cambridge, Mass.: MIT Press, 1965.

Levine, Adeline. *Love Canal: Science, Politics and People.* Lexington, Mass.: Lexington Books, 1982.

Lifton, Robert. *Death in Life: Survivors of Hiroshima.* New York: Random House, 1968.

———. *The Broken Connection: On Death and the Continuity of Life.* New York: Simon and Schuster, 1979.

Sorenson, John, et al. *Impacts of Hazardous Technology: The Psychosocial Effects of Restarting TMI-1.* Albany, N.Y.: State University of New York Press, 1987.

Sowder, Barbara, ed. *Disasters and Mental Health: Selected Contemporary Perspectives.* Rockville, Md.: National Institute of Mental Health, 1987.

Vyner, Henry. *Invisible Trauma: The Psychosocial Effects of the Invisible Environmental Contaminants.* Lexington, Mass.: Lexington Books, 1988.

Wolfenstein, Martha. *Disaster: A Psychological Essay.* New York: Arno Press, 1957, 1977.

THE FUTURE: WHAT TO THINK AND DO

Anderson, Walter Truett. *To Govern Evolution: Further Adventures of the Human Animal.* New York: Harcourt Brace Jovanovich, 1987.

Bateson, Gregory. *Steps to an Ecology of Mind.* New York: Ballantine, 1972.

Berger, John. *Restoring the Earth: How Americans Are Working to Renew Our Damaged Environment,* New York: Doubleday, 1987.

Bookchin, Murray. *The Evolution of Freedom.* Palo Alto, Calif.: Cheshire Books, 1982.

Bower, John. *The Healthy House.* London: Lyle Stuart, 1989.

Brown, Lester. *Building a Sustainable Society.* New York: Norton, 1981.

———, et al., *State of the World 1984: Progress Toward a Sustainable Society.* New York: Norton, 1984.

———, et al., *State of the World 1985: Progress Toward a Sustainable Society.* New York: Norton, 1985.

———, et al. *State of the World 1986: Progress Toward a Sustainable Society.* New York: Norton, 1986.

————, et al. *State of the World 1987: Progress Toward a Sustainable Society.* New York: Norton, 1987.

————, et al. *State of the World 1988: Progress Toward a Sustainable Society.* New York: Norton, 1988.

————, et al. *State of the World 1989: Progress Toward a Sustainable Society.* New York: Norton, 1989.

Capra, Fritjof. *The Turning Point.* New York: Simon and Schuster, 1982.

Dadd, Debra. *The Nontoxic Home.* Los Angeles: Tarcher, 1986.

————. *Nontoxic and Natural.* Los Angeles: Tarcher, 1984.

Duvall, Bill, and George Sessions. *Deep Ecology: Living as if Nature Mattered.* Salt Lake City: Peregrine Smith Books, 1985.

Elgin, Duane. *Voluntary Simplicity.* New York: William Morrow, 1981.

Fight to Win On Hazardous Waste: A Leader's Manual. Arlington VA: Citizens Clearinghouse On Hazardous Waste, 1982.

Good, Clint, and Debra Dadd. *Healthful Houses: How to Design and Build Your Own.* Bethesda, Maryland: Guaranty Press, 1988.

Harman, Willis. *An Incomplete Guide to the Future.* Stanford, Calif.: Stanford Alumni Association, 1976.

Henderson, Hazel. *Creating Alternative Futures.* New York, Putnam, 1978.

————. *The Politics of the Solar Age.* New York: Anchor/Doubleday, 1981.

Lovins, Amory. *Soft Energy Paths: Towards a Durable Peace.* New York: Ballinger, 1977.

Ott, John. *Health and Light.* New York: Pocket Books, 1973.

Rifkin, Jeremy. *Declaration of a Heretic.* New York: Routledge, Chapman and Hall, 1985.

Sherry, Susan, et al. *High Tech and Toxics: A Guide for Local Communities.* Washington, D.C.: Conference on Alternative State and Local Policies, 1985.

Toxics Liability Workbook. Washington, D.C.: Environmental Action, 1988.

Van der Ryn, Sim, and Peter Calthorpe. *Sustainable Communities: A New Design Synthesis for Cities, Suburbs, and Towns.* San Francisco: Sierra Club Books, 1986.

Walsh, Roger. *Staying Alive: The Psychology of Human Survival.* Boston: Shambhala, 1984.

Winner, Langdon. *The Whale and the Reactor: A Search for Limits in an Age of High Technology.* Chicago: University of Chicago Press, 1986.

APPENDIX

THE STUDY: PSYCHOLOGICAL EFFECTS OF TECHNOLOGY-INDUCED DISEASES

The research reported in this book is based on a case study undertaken between 1986 and 1988, focusing on the psychological responses of technology survivors. These are people who have become ill through exposure to or use of modern, health-threatening technologies. They sustain technology-induced diseases. The study addresses the overall psychological issues these people face and the varied responses they develop for coping.

I found interviewees in a number of ways: (1) by obtaining lists from support and political action groups, like the National Association of Radiation Survivors and DES Action, and then randomly selecting potential interviewees; (2) by running advertisements in newspapers requesting volunteers for interviews; and (3) through contacts provided by friends and colleagues.

Before interviewing candidates, I requested medical and historical documentation of the connection between their illness and an offending technology. Due to a current lack of scientific proof, however, final evaluation of this connection had to be subjective in several cases. Where the connection was particularly unverifiable, either the person was screened from participation or the experience was used to support theories concerning uncertainty (Chapter 10).

In the end, forty-six people participated. Each was interviewed in depth once or twice, for a total time span ranging from two to five hours.

All of the forty-six were Americans. Thirty-two were women, fourteen men. Exposures to health-threatening technologies took place in the following locations: the South Pacific, two cases; Japan, three; California, thirteen; Colorado, two; Iowa, one; Massachusetts, five; Michigan, one; Minnesota, one; Nevada, one; New Mexico, six; New York, three; Ohio, two; Oklahoma, one; Texas, two; Utah, two; Washington State, two; Washington, D.C., one. (The total exceeds forty-six due to multiple exposures by several people in different locations.)

In racial and ethnic background, the participants included seven Hispanics, three African-Americans, thirty-three Caucasians, two Asians, and one Native American.

At the time of the interviews, four participants were under the age of twenty-five. Thirteen were between twenty-six and forty years old. Twenty-one were between forty-one and sixty. Eight were over sixty.

Ten people had incomes under $10,000 at the time of the interviews. Twelve were making between $10,000 and $25,000. Sixteen had incomes between $25,000 and $50,000, while four were making over $50,000. (The total number of incomes is smaller than forty-six because several interviewees did not reveal this information.)

Two participants reported their illnesses had been brief. Seventeen indicated that theirs had been treated and healed. Twenty-six reported chronic diseases, eleven of those terminal. In this category I count illnesses requiring irreversible surgery, such as hysterectomy or removal of a lung, as chronic. Also chronic are genetic defects that are passed on to subsequent generations.

Affecting technologies included the following. *Occupational technologies:* asbestos, five cases; ionizing radiation, one; pesticides, one; other toxic chemicals, seven. *Military technology:* ionizing radiation, six. *Consumer technologies:* nonionizing radiation, one; pesticides, two; other toxic chemicals, one. *Residential technologies:* ionizing radiation, two; pesticides, four; other toxic chemicals, three. *Medical technologies:* intrauterine devices, nine; antibiotics in overdose, four; oral contraceptives and estrogen replacement therapy, four; diethylstilbestrol, three; amalgam dental work, one. (The total exceeds forty-six due to multiple exposures by a number of people.)

The two- to five-hour interview uniformly included the following questions:

1. What technology do you believe caused your illness? Please describe your exposure to/use of this technology and subsequent events relating to your health.
2. At the time of exposure to/use of this technology, had you ever heard of its potential danger to health?
3. How did you discover the connection between your illness and exposure/use? How long was it from the time of exposure/use until you discovered the connection? How did you feel when you made the discovery?
4. According to medical professionals you have seen, what is your diagnosis and prognosis? In your view, do these professionals understand your illness? How do you feel about your medical condition?
5. How has your illness affected your life economically?
6. Have you attempted to get compensation for medical expenses, lost

wages, or suffering? With what results? How do you feel about this process and the results?

7. Are you a member of a group of people who have had a similar experience? If yes, why did you join? What activities are you engaged in with the group? How do you feel about being part of it?

8. How has your experience affected your family and immediate community?

9. What has the worst part of the experience been for you?

10. What resources have you found or invented to help you cope?

11. What have you learned from your experience with technology and illness?

12. What do you think should be done on a social, political, or cultural level to address the kind of technological event you have described?

13. What is your vision of a society that uses technology in safe, beneficial ways?

14. If you could tell the world one thing about your experience, what would it be?

15. Is there anything else you would like to say?

16. How was it for you to talk about this subject today?

INDEX

275

AVA, *see* Asbestos Victims of
America
Ayurvedic medicine, 158

Baca, Irene, 146, 149, 161, 204
chemical exposure of, 28, 72, 93–94
compensation received by, 108–109
confidence lost by, 62, 70
consequences of illness of, 67
on corporate responsibility, 202,
208
medical profession and, 72, 94
occupational innocence of, 104
bankruptcy, Chapter 11 declarations
of, 109–110
Baum, Andrew, 47
Beck, Judy, 76
Becker, Ernest, 79
Beinfield, Harriet, 26, 64, 88–89,
169
as acupuncturist, 177
birth defects in son of, 64, 68, 88,
96, 163, 169
emotional state of, 96
faith of, in self, 168
on progress, 136, 205–206
sense of isolation of, 68–69
beliefs, collectively held, 135–146
computers in, 144–146
mechanization of life as, 142–144
progress as, 136–137
technical solutions as, 137–140
technological control as, 140–142
benzene, 20, 27, 35, 36, 84, 98
health effects of, 51, 52, 55–56, 89,
95
Berner, Bruce, 107
Berning, Betsy:
antibiotics and, 28, 74
Dalkon Shield and, 28, 47, 165
on self-help, 150, 152–153, 168
Bikini Island, as nuclear test site,
45–46, 186, 197
birth control pills, 21
advertising for, 123
Beinfield and, 26
Glendinning and, 15–16, 21, 49
Gould and, 27, 79, 92
Griffin and, 27
health effects of, 15, 57, 123
Jans and, 28, 105–106
Landau and, 26

number of users of, 18
Tavich and, 27, 47–48, 69
birth defects, 17, 48, 55, 56, 57, 64,
67, 68, 88, 96, 163, 169
Bliss, Shepherd, 173
blue baby syndrome, 36
bone disease, 53, 66
Borel, Clarence, 102
Bradley, David, 186
brain damage, 34, 35, 203
brain tumors, 32, 71,. 124
Brim, Raymond, 32
Bruen, Bliss, 28, 54, 132, 203
loss of trust by, 84
on political change, 204, 208
self-education of, 156–157
Buen Samaritano Methodist Church,
El, 36
Buffalo Creek, mining disaster at, 59,
68, 84
Buford, Anne, 111
Bulman-Janoff, Ronnie, 24
Bush, George, 32
Butrico, Frank, 91

California, University of, radiation
lawsuit against, 108
Calvert, Loran, 207, 212
asbestos and, 27, 54, 130
on helping others, 175
suicidal impulses of, 65
Campaign Against Toxic Hazards,
186
canaries, used in coal mines, 59–60
cancer, 45, 53
causes of, 31–32, 34, 36, 45, 48, 50,
57, 58, 92, 112, 118, 125
cervical, 47, 57
DES and, 48
increase in, 56
malathion and, 112
Cancer Alley, 118
Candelaria, Ricardo, 156, 166, 205
accused of unpatriotic behavior, 69
community involvement of, 173–174
DVA and, 75, 76–77, 207
exposed to nuclear radiation, 28, 66,
81–82
loss of purpose felt by, 79
sterility of, 81–82
candida, *see* systemic candidiasis
carbon monoxide, 20

276

277

278

Totem and Taboo (Freud), 115
Toxic Substance Strategy, U.S., 56
toxic waste sites:
    EPA Superfund cleanup program
      for, 111
    numbers of, 20, 119, 195
"Traumatic Neurosis" (Kardiner), 41
20/20, 76
2,4-D, 17, 88
2,4,5-T, 17, 88
2,3,7,8-tetrachlorodibenzo-para-dioxin
    (TCDD), 90, 93

Union of Technology Survivors,
    proposal for, 212–213
United Nations Conference on
    Women, 175
uranium, 18, 103–104

vaginal infections, 15
vaginitis, 47
VDTs, *see* video display terminals,
    radiation from
vegetable gums, 58
Vermonters Organized for Cleanup,
    187
Veterans Administration, 75, 91, 207
Veterans Affairs Department, U.S.
    (DVA), Agent Orange claims
    and, 75–77, 106, 108
victimization, 59–78, 149
    financial aspect of, 66–67
    hidden numbers and, 117–120
    institutions and, 77–78
    legal system and, 99–113
    loss of identity and, 79–86
    medical profession and, 70–77, 83,
      84, 119, 124, 200, 201–202, 204
    moving beyond, 149–162
    peer group pressure and, 105
    psychological process of, 59–65, 200
    social aspect of, 68–70
    uncertainty of, 87–97
video display terminals (VDTs),
    radiation from, 19, 57, 117,
    118–119, 146

vinyl chloride, 20, 58
vitamin C, 33, 160
vitamins, 160
Vyner, Henry, 75
    on new diseases, 73–74, 93
    on patient-doctor relationship,
      70
    psychological work of, with
      radiation victims, 21, 68, 92, 95,
      152, 165

Waelder, Robert, 70
Wagner, Michael, 53
Waldsterban, 192
Wallace, Carey, 27
Warner Amendment, 108, 185
Warwick Against Radium Dumping,
    162
Waters, Bob, 171
weed killer, 17, 27
Wescott, Joan, 199, 200
West County Toxics Coalition, 187
Whiteside, Thomas, 48
Wilkinson, Christopher, 128
Winner, Langdon, 13, 22, 23, 141,
    144
Withering Rain (Whiteside), 48
Wolfenstein, Martha, 124, 162, 166,
    171
Wolman, Carol, 129
*Woman and Nature: The Roaring
    Inside Her* (Griffin), 189
Woolf, Jane, 193
    benzene contamination and, 27,
      89
    fear felt by, 52
    optimism of, 153, 155, 166, 207
    uncertainty felt by, 95
workers' compensation, unfairness of,
    108

X rays, 54
    number of pregnant women exposed
      to, 19
    in prison experiment, 103
xylene, 27, 162